Understanding Big Data

About the Authors

Paul C. Zikopoulos, B.A., M.B.A., is the Director of Technical Professionals for IBM Software Group's Information Management division and additionally leads the World Wide Database Competitive and Big Data SWAT teams. Paul is an internationally recognized award-winning writer and speaker with more than 18 years of experience in Information Management. Paul has written more than 350 magazine articles and 14 books on database technologies, including *DB2 pureScale: Risk Free Agile Scaling* (McGraw-Hill, 2010); *Break Free with DB2 9.7: A Tour of Cost-Slashing New Features* (McGraw-Hill, 2010); *Information on Demand: Introduction to DB2 9.5 New Features* (McGraw-Hill, 2007); *DB2 Fundamentals Certification for Dummies* (For Dummies, 2001); *DB2 for Windows for Dummies* (For Dummies, 2001), and more. Paul is a DB2 Certified Advanced Technical Expert (DRDA and Clusters) and a DB2 Certified Solutions Expert (BI and DBA). In his spare time, he enjoys all sorts of sporting activities, including running with his dog, Chachi; avoiding punches in his MMA training; trying to figure out why his golf handicap has unexplainably decided to go up; and trying to understand the world according to Chloë, his daughter. You can reach him at paulz_ibm@msn.com. Also, keep up with Paul's take on Big Data by following him on Twitter @BigData_paulz.

Chris Eaton, B.Sc., is a worldwide technical specialist for IBM's Information Management products focused on Database Technology, Big Data, and Workload Optimization. Chris has been working with DB2 on the Linux, UNIX, and Windows platform for more than 19 years in numerous roles, from support, to development, to product management. Chris has spent his career listening to clients and working to make DB2 a better product. He is the author of several books in the data management space, including *The High Availability Guide to DB2* (IBM Press, 2004), *IBM DB2 9 New Features* (McGraw-Hill, 2007), and *Break Free with DB2 9.7: A Tour of Cost-Slashing New Features* (McGraw-Hill, 2010). Chris is also an international award-winning speaker, having presented at data management conferences across the globe, and he has one of the most popular DB2 blogs located on IT Toolbox at http://it.toolbox.com/blogs/db2luw.

Dirk deRoos, B.Sc., B.A., is a member of the IBM World-Wide Technical Sales Team, specializing in the IBM Big Data Platform. Dirk joined IBM 11 years ago and previously worked in the Toronto DB2 Development lab as its

Information Architect. Dirk has a Bachelor's degree in Computer Science and a Bachelor of Arts degree (Honors in English) from the University of New Brunswick.

Thomas Deutsch, B.A, M.B.A., serves as a Program Director in IBM's Big Data business. Tom has spent the last couple of years helping customers with Apache Hadoop, identifying architecture fit, and managing early stage projects in multiple customer engagements. He played a formative role in the transition of Hadoop-based technologies from IBM Research to IBM Software Group, and he continues to be involved with IBM Research Big Data activities and the transition of research to commercial products. Prior to this role, Tom worked in the CTO office's Information Management division. In that role, Tom worked with a team focused on emerging technologies and helped customers adopt IBM's innovative Enterprise Mashups and Cloud offerings. Tom came to IBM through the FileNet acquisition, where he had responsibility for FileNet's flagship Content Management product and spearheaded FileNet product initiatives with other IBM software segments including the Lotus and InfoSphere segments. With more than 20 years in the industry and a veteran of two startups, Tom is an expert on the technical, strategic, and business information management issues facing the enterprise today. Tom earned a Bachelor's degree from the Fordham University in New York and an MBA from the Maryland University College.

George Lapis, MS CS, is a Big Data Solutions Architect at IBM's Silicon Valley Research and Development Lab. He has worked in the database software area for more than 30 years. He was a founding member of R* and Starburst research projects at IBM's Almaden Research Center in Silicon Valley, as well as a member of the compiler development team for several releases of DB2. His expertise lies mostly in compiler technology and implementation. About ten years ago, George moved from research to development, where he led the compiler development team in his current lab location, specifically working on the development of DB2's SQL/XML and XQuery capabilities. George also spent several years in a customer enablement role for the Optim Database toolset and more recently in IBM's Big Data business. In his current role, George is leading the tools development team for IBM's InfoSphere BigInsights platform. George has co-authored several database patents and has contributed to numerous papers. He's a certified DB2 DBA and Hadoop Administrator.

About the Technical Editor

Steven Sit, B.Sc., MS, is a Program Director in IBM's Silicon Valley Research and Development Lab where the IBM's Big Data platform is developed and engineered. Steven and his team help IBM's customers and partners evaluate, prototype, and implement Big Data solutions as well as build Big Data deployment patterns. For the past 17 years, Steven has held key positions in a number of IBM projects, including business intelligence, database tooling, and text search. Steven holds a Bachelor's degree in Computer Science (University of Western Ontario) and a Masters of Computer Science degree (Syracuse University).

Understanding Big Data
Analytics for Enterprise Class Hadoop and Streaming Data

Paul C. Zikopoulos
Chris Eaton
Dirk deRoos
Thomas Deutsch
George Lapis

New York   Chicago   San Francisco
Lisbon   London   Madrid   Mexico City
Milan   New Delhi   San Juan
Seoul   Singapore   Sydney   Toronto

McGraw-Hill books are available at special quantity discounts to use as premiums and sales promotions, or for use in corporate training programs. To contact a representative, please e-mail us at bulksales@mcgraw-hill.com.

Understanding Big Data: Analytics for Enterprise Class Hadoop and Streaming Data

The contents of this book represent those features that may or may not be available in the current release of any products mentioned within this book despite what the book may say. IBM reserves the right to include or exclude any functionality mentioned in this book for the current release of InfoSphere Streams or InfoSphere BigInsights, or a subsequent release. In addition, any performance claims made in this article are not official communications by IBM; rather the results observed by the authors in unaudited testing. The views expressed in this article are those of the authors and not necessarily those of IBM Corporation.

1 2 3 4 5 6 7 8 9 0   DOC DOC   10 9 8 7 6 5 4 3 2 1

ISBN   978-0-07-179053-6

MHID   0-07-179053-5

Sponsoring Editor	**Copy Editor**	**Illustration**
Paul Carlstroem	Lisa Theobald	Cenveo Publisher Services
Editorial Supervisor	**Proofreader**	**Art Director, Cover**
Patty Mon	Paul Tyler	Jeff Weeks
Project Manager	**Production Supervisor**	
Sheena Uprety, Cenveo Publisher Services	George Anderson	
Acquisitions Coordinator	**Composition**	
Stephanie Evans	Cenveo Publisher Services	

My fifteenth book in my eighteenth year at IBM—it's hard to believe so much time has passed and Information Management technology has become not just my career, but somewhat of a hobby (insert image of Chloe reading this in a couple of years once she learns the universal "loser" gesture).

I often dedicate my books to people in my life: This book I actually want to dedicate to the company in my life that turned 100 years old on August 12, 2011: IBM. In this day and age of fluid careers, the U.S. Department of Labor has remarked that the average learner will have 10 to 14 jobs by the time they are 38; 1 in 4 workers have been with their employer less than a year; and 1 in 2 workers have been with their employer less than 5 years. Sometimes I get asked about my 18-year tenure at IBM in a tone of disbelief for my generation. In my 18 years at IBM, I've had the honor to learn and participate in the latest technologies, marketing, sales, technical sales, writing, usability design, development, partner programs, channels, education, support, services, public speaking, competitive analysis, and always learning more. IBM has always been a place that nurtures excellence and opportunity for those thirsty to make a difference, and I've got a thirst not yet quenched. IBM deeply encourages learning from others—and I often wonder if other people feel like they won the lottery with a mentoring team (Martin Wildberger, Bob Piciano, Dale Rebhorn, and Alyse Passurelli) like the one I have. Thanks to IBM for providing an endless cup of opportunity and learning experiences.

Finally, to my two gals, whose spirits always run through my soul: Grace Madeleine Zikopoulos and Chloë Alyse Zikopoulos.

—Paul Zikopoulos

This is the fourth book that I have authored, and every time I dedicate the book to my wife and family. Well this is no exception, because it's their support that makes this all possible, as anyone who has ever spent hours and hours of their own personal time writing a book can attest to.

To my wife, Teresa, who is always supporting me 100 percent in all that I do, including crazy ideas like writing a book. She knows full well how much time it takes to write a book since she is a real author herself and yet she still supported me when I told her I was going to write this book (you are a saint). And to Riley and Sophia, who are now old enough to read one of my books (not that they are really interested in any of this stuff since they are both under ten). Daddy is finished writing his book so let's go outside and play.

—Chris Eaton

I'd like to thank Sandra, Erik, and Anna for supporting me, and giving me the time to do this. Also, thanks to Paul for making this book happen and asking me to contribute.

—Dirk deRoos

I would like to thank my ridiculously supportive wife and put in writing for Lauren and William that yes, I will finally take them to Disneyland again now that this is published. I'd also like to thank Anant Jhingran for both the coaching and opportunities he has entrusted in me.

—Thomas Deutsch

"If you love what you do, you will never work a day in your life." I dedicate this book to all my colleagues at IBM that I worked with over the years who helped me learn and grow and have made this saying come true for me.

—George Lapis

Thanks to my IBM colleagues in Big Data Research and Development for the exciting technologies I get to work on every day. I also want to thank Paul for the opportunity to contribute to this book. Last but not least, and most importantly, for my wife, Amy, and my twins, Tiffany and Ronald, thank you for everything you do, the joy you bring, and for supporting the time it took to work on this book.

—Steven Sit

CONTENTS AT A GLANCE

CONTENTS

PART I
Big Data: From the Business Perspective

FOREWORD

Executive Letter from Rob Thomas

There's an old story about two men working on a railroad track many years back. As they are laying track in the heat of the day, a person drives by in a car and rolls down the window (not enough to let the air conditioning out, but enough to be heard). He yells, "Tom, is that you?" Tom, one of the men working on the track, replies, "Chris, it's great to see you! It must have been 20 years…How are you?" They continue the conversation and Chris eventually drives off. When he leaves, another worker turns to Tom and says, "I know that was the owner of the railroad and he's worth nearly a billion dollars. How do you know him?" Tom replies, "Chris and I started working on the railroad, laying track, on the same day 20 years ago. The only difference between Chris and me is that I came to work for $1.25/hour and he came to work for the railroad."

*****

Perspective. Aspiration. Ambition. These are the attributes that separate those who come to work for a paycheck versus those who come to work to change the world. The coming of the Big Data Era is a chance for everyone in the technology world to decide into which camp they fall, as this era will bring the biggest opportunity for companies and individuals in technology since the dawn of the Internet.

Let's step back for a moment and look at how the technology world has changed since the turn of the century:

- 80 percent of the world's information is unstructured.

- Unstructured information is growing at 15 times the rate of structured information.

- Raw computational power is growing at such an enormous rate that today's off-the-shelf commodity box is starting to display the power that a supercomputer showed half a decade ago.

- Access to information has been democratized: it is (or should be) available for all.

This is the new normal. These aspects alone will demand a change in our approach to solving information-based problems. Does this mean that our investments in the past decade are lost or irrelevant? Of course not! We will still need relational data stores and warehouses, and that footprint will continue to expand. However, we will need to augment those traditional approaches with technology that will allow enterprises to benefit from the Big Data Era.

The Big Data Era will be led by the individuals and companies that deliver a platform suitable for the new normal—a platform consisting of exploration and development toolsets, visualization techniques, search and discovery, native text analytics, machine learning, and enterprise stability and security, among other aspects. Many will talk about this, few will deliver.

I'm participating here because I know we can change the technology world, and *that's* much more satisfying than $1.25/hour. Welcome to the Big Data Era.

Rob Thomas
IBM Vice President, Business Development

Executive Letter from Anjul Bhambhri

It was in the 1970s when the first prototype of a relational database system, *System R*, was created in the Almaden Research labs in San Jose. System R sowed the seeds for the most common way of dealing with data structured in relational form, called SQL; you'll recognize it as a key contribution to the development of products such as DB2, Oracle, SQL/DS, ALLBASE, and Non-Stop SQL, among others. In combination with the explosion of computing power across mainframes, midframes, and personal desktops, databases have become a ubiquitous way of collecting and storing data. In fact, their proliferation led to the creation of a discipline around "warehousing" the data, such that it was easier to manage and correlate data from multiple databases in a uniform fashion. It's also led to the creation of vertical slices of these warehouses into data marts for faster decisions that are tightly associated with specific lines of business needs. These developments, over a short period of ten years in the 1990s, made the IT department a key competitive differentiator for every business venture. Thousands of applications were born—some horizontal across industries, some specific to domains such as purchasing, shipping, transportation, and more. Codenames such as ERP (Enterprise Resource Planning), SCM (Supply Chain Management), and others became commonplace.

By the late 1990s, inevitably, different portions of an organization used different data management systems to store and search their critical data, leading to federated database engines (under the IBM codename *Garlic*). Then, in 2001, came the era of XML. The DB2 pureXML technology offers sophisticated capabilities to store, process, and manage XML data in its native hierarchical format. Although XML allowed a flexible schema and ease of portability as key advantages, the widespread use of e-mail, accumulation of back office content, and other technologies led to the demand for content management systems and the era of analyzing unstructured and semistructured data in enterprises was born. Today, the advent of the Internet, coupled with complete democratization of content creation and distribution in multiple formats, has led to the explosion of all types of data. Data is now not only big, both in terms of volume and variety, but it has a velocity component to it as well. The ability for us to glean the nuggets of information embedded in such a cacophony of data, at precisely the time of need, makes it very exciting. We are sitting at the cusp of another evolution, popularly called as *Big Data*.

At IBM, our mission is to help our clients achieve their business objectives through technological innovations, and we've being doing it for a century as of 2011. During the last five decades, IBM has invented technologies and delivered multiple platforms to meet the evolving data management challenges of our customers. IBM invented the relational database more than 30 years ago, which has become an industry standard with multiple products provided by IBM alone (for example, DB2, Informix, Solid DB, and others). Relational databases have further specialized into multidimensional data warehouses, with highly parallel data servers, a breadth of dedicated appliances (such as Netezza or the Smart Analytics System), as well as analysis and reporting tools (such as SPSS or Cognos).

Across industries and sectors (consumer goods, financial services, government, insurance, telecommunications, and more), companies are assessing how to manage the volume, variety, and velocity of their untapped information in an effort to find ways to make better decisions about their business. This explosion of data comes from a variety of data sources such as sensors, smart devices, social media, billions of Internet and smartphone users, and more. This is data that arrives in massive quantities in its earliest and most primitive form.

Organizations seeking to find a better way, which differentiates them from their competitors, want to tap into the wealth of information hidden in this explosion of data around them to improve their competitiveness, efficiency, insight, profitability, and more. These organizations recognize the value delivered by analyzing all their data (structured, semistructured, and unstructured) coming from a myriad of internal and external sources. This is the realm of "Big Data." While many companies appreciate that the best Big Data solutions work across functional groups touching many positions, few corporations have figured out how to proceed. The challenge for the enterprise is to have a data platform that leverages these large volumes of data to derive timely insight, while preserving their existing investments in Information Management. In reality, the best Big Data solutions will also help organizations to know their customer better than ever before.

To address these business needs, this book explores key case studies of how people and companies have approached this modern problem. The book diligently describes the challenges of harnessing Big Data and provides examples of Big Data solutions that deliver tangible business benefits.

I would like to thank Paul, George, Tom, and Dirk for writing this book. They are an outstanding group whose dedication to our clients is unmatched. Behind them is the Big Data development team, who continually overcomes the challenges of our decade. I get to work with an outstanding group of people who are passionate about our customers' success, dedicated to their work, and are continually innovating. It is a privilege to work with them.

Thank you, and enjoy the book.

Anjul Bhambhri
IBM Vice President, Big Data Development

ACKNOWLEDGMENTS

Collectively, we want to thank the following people, without whom this book would not have been possible: Shivakumar (Shiv) Vaithyanathan, Roger Rea, Robert Uleman, James R. Giles, Kevin Foster, Ari Valtanen, Asha Marsh, Nagui Halim, Tina Chen, Cindy Saracco, Vijay R. Bommireddipalli, Stewart Tate, Gary Robinson, Rafael Coss, Anshul Dawra, Andrey Balmin, Manny Corniel, Richard Hale, Bruce Brown, Mike Brule, Jing Wei Liu, Atsushi Tsuchiya, Mark Samson, Douglas McGarrie, Wolfgang Nimfuehr, Richard Hennessy, Daniel Dubriwny, our Research teams, and all the others in our business who make personal sacrifices day in and day out to bring you the IBM Big Data platform.

Rob Thomas and Anjul Bhambhri deserve a special mention because their passion is contagious—thanks to both of you.

We especially want to give a heartfelt thanks to our terrific Distinguished Engineer (DE), Steve Brodsky, and the two lead Senior Technical Staff Members (STSMs) on BigInsights: Shankar Venkataraman, and Bert Van der Linden; without their dedication and efforts, this book would not be possible. IBM is an amazing place to work, and becomes unparalleled when you get to work, day in and day out, beside the kind of brainpower these guys have and their good natured willingness to share it and make us all smarter. We would also be remiss not to thank Steven Sit, who at the last minute came in to be our technical editor (and part-time researcher, though we failed to tell him that when we approached him with the role).

We want to thank (although at times we cursed) Susan Visser and Linda Currie for getting the book in place; an idea is an idea, but it takes people like this to help get that idea off of a Buffalo wings stained napkin and into your hands without the mess. Our editing team—Sheena Uprety, Patty Mon, Paul Tyler, and Lisa Theobald—all played a key role behind the scenes and we want to extend our thanks for that. Thanks, to our McGraw-Hill guru, Paul Carlstroem—there is a reason why we specifically wanted to work with you. (By the way, the extra week came in handy!)

Finally, to Linda Snow for taking time away from her precious Philadelphia Eagles on "Don't talk to me it's football" days and Wendy Lucas, for taking the time out of their busy lives to give the book a read and keep us on the right track. You two are great colleagues and our clients are lucky to have you in the field, working on their success with the passion you each bring to our business.

ABOUT THIS BOOK

This book's authoring team is well seasoned in traditional database technologies, and although we all have different backgrounds and experiences at IBM, we all recognize one thing: Big Data is an inflection point when it comes to information technologies: in short, Big Data *is* a Big Deal! In fact, Big Data is going to change the way you do things in the future, how you gain insight, and how you make decisions (this change isn't going to be a replacement for the way things are done today, but rather a highly valued and much anticipated extension).

Recognizing this inflection point, we decided to spend our recent careers submersing ourselves in Big Data technologies and figured this book was a great way to get you caught up fast if it's all new to you. We hope to show you the unique things IBM is doing to embrace open source Big Data technologies, such as Hadoop, and extending it into an enterprise ready *Big Data Platform*. The IBM Big Data platform uses Hadoop as its core (there is no forking of the Apache Hadoop code and BigInsights always maintains backwards compatibility with Hadoop) and marries that to enterprise capabilities provided by a proven visionary technology leader that understands the benefits a platform can provide. IBM infuses its extensive text analytics and machine learning intellectual properties into such a platform, hardens it with an industry tried, tested, and true enterprise-grade file system, provides enterprise integration, security, and more. We are certain you can imagine the possibilities. IBM's goal here isn't to get you a running Hadoop cluster—that's something we do along the path; rather, it's to give you a new way to gain insight into vast amounts of data that you haven't easily been able to tap into before; that is, until a technology like Hadoop got teamed with an analytics leader like IBM. In short, IBM's goal is to help you meet your analytics challenges and give you a platform to create an end-to-end solution.

Of course, the easier a platform is to use, the better the return on investment (ROI) is going to be. When you look at IBM's Big Data platform, you can see all kinds of areas where IBM is flattening the time to analysis curve with Hadoop. We can compare it to the cars we drive today. At one end of the spectrum, a standard transmission can deliver benefits (gas savings, engine braking, and acceleration) but requires a fair amount of complexity to learn (think about the first time you drove "stick"). At the other end of the spectrum, an automatic transmission doesn't give you granular control when you need it, but is far easier to operate. IBM's Big Data platform has morphed itself a Porsche-like Doppelkupplung transmission—you can use it in automatic mode to get up and running quickly with text analysis for data in motion and data-at-rest, and you can take control and extend or roll your own analytics to deliver localized capability as required. Either way, IBM will get you to the end goal faster than anyone.

When IBM introduced the world to what's possible in a Smarter Planet a number of years ago, the company recognized that the world had become *instrumented*. The transistor has become the basic building block of the digital age. Today, an average car includes more than a million lines of code; there are 3 million lines of code tracking your checked baggage (with that kind of effort, it's hard to believe that our bags get lost as often as they do); and more than a billion lines of code are included in the workings of the latest Airbus plane.

Quite simply (and shockingly), we now live in a world that has more than a billion transistors per human, each one costing one ten-millionth of a cent; a world with more than 4 billion mobile phone subscribers and about 30 billion radio frequency identification (RFID) tags produced globally within two years. These sensors all generate data across entire ecosystems (supply chains, healthcare facilities, networks, cities, natural systems such as waterways, and so on); some have neat and tidy data structures, and some don't. One thing these instrumented devices have in common is that they all generate data, and that data holds an opportunity cost. Sadly, due to its voluminous and non-uniform nature, and the costs associated with it, much of this data is simply thrown away or not persisted for any meaningful amount of time, delegated to "noise" status because of a lack of efficient mechanisms to derive value from it.

A Smarter Planet, by a natural extension of being instrumented, is *interconnected*. Sure, there are almost 2 billion people using the Internet, but think about all those instrumented devices having the ability to talk with each other. Extend this to the prospect of a trillion connected and intelligent objects ranging from bridges, cars, appliances, cameras, smartphones, roadways, pipelines, livestock, and even milk containers and you get the point: the amount of information produced by the interaction of all those data generating and measuring devices is unprecedented, but so, too, are the challenges and potential opportunities.

Finally, our Smarter Planet has become *intelligent*. New computing models can handle the proliferation of end user devices, sensors, and actuators, connecting them with back-end systems. When combined with advanced analytics, the right platform can turn mountains of data into intelligence that can be translated into action, turning our systems into intelligent processes. What this all means is that digital and physical infrastructures of the world have arguably converged. There's computational power to be found in things we wouldn't traditionally recognize as computers, and included in this is the freeform opportunity to share with the world what you think about pretty much anything. Indeed, almost anything—any person, object, process, or service, for any organization, large or small—can become digitally aware and networked. With so much technology and networking abundantly available, we have to find cost-efficient ways to gain insight from all this accumulating data.

A number of years ago, IBM introduced business and leaders to a Smarter Planet: directional thought leadership that redefined how we think about technology and its problem-solving capabilities. It's interesting to see just how much foresight IBM had when it defined a Smarter Planet, because all of those principles seem to foreshadow the need for a Big Data platform.

Big Data has many use cases; our guess is that we'll find it to be a ubiquitous data analysis technology in the coming years. If you're trying to get a handle on brand sentiment, you finally have a cost-efficient and capable framework to measure cultural decay rates, opinions, and more. Viral marketing is nothing new. After all, one of its earliest practitioners was Pyotr Smirnov (yes, the vodka guy). Smirnov pioneered charcoal filtration, and to get his message out, he'd hire people to drink his vodka at establishments everywhere and boisterously remark as to its taste and the technology behind it. Of course, a Smarter Planet takes viral to a whole new level, and a Big Data platform provides a transformational information management platform that allows you to gain insight into its effectiveness.

Big Data technology can be applied to log analysis for critical insight into the technical underpinnings of your business infrastructure that before had to be discarded because of the amount of something we call *Data Exhaust*. If your platform gave you the ability to easily classify this valuable data into noise and signals, it would make for streamlined problem resolution and preventative processes to keep things running smoothly. A Big

Data platform can deliver ground-breaking capability when it comes to fraud detection algorithms and risk modeling with expanded models that are built on more and more identified causal attributes, with more and more history—the uses are almost limitless.

This book is organized into two parts. *Part I*—Big Data: From the Business Perspective focuses on the *who* (it all starts with a kid's stuffed toy—read the book if that piqued your curiosity), *what*, *where*, *why*, and *when* (it's not too late, but if you're in the Information Management game, you can't afford to delay any longer) of Big Data. Part I is comprised of three chapters.

Chapter 1 talks about the three defining characteristics of Big Data: *volume* (the growth and run rates of data), *variety* (the kinds of data such as sensor logs, microblogs—think Twitter and Facebook—and more), and *velocity* (the source speed of data flowing into your enterprise). You're going to hear these three terms used a lot when it comes to Big Data discussions by IBM, so we'll often refer to them as *"the 3 Vs"*, or "V^3" throughout this book and in our speaking engagements. With a firm definition of the characteristics of Big Data you'll be all set to understand the concepts, use cases, and reasons for the technologies outlined in the remainder of this book. For example, think of a typical day, and focus on the 30 minutes (or so) it takes for one of us to drive into one of the IBM labs: in the amount of time it takes to complete this trip, we've generated and have been subjected to an incredible number of Big Data events.

From taking your smartphone out of its holster (yes, that's a recorded event for your phone), to paying road tolls, to the bridge one of us drives over, to changing an XM radio station, to experiencing a media impression, to checking e-mails (not while driving of course), to badging into the office, to pressing *Like* on an interesting Facebook post, we're continually part of Big Data's V^3. By the way, as we've implied earlier, you don't have to breathe oxygen to generate V^3 data. Traffic systems, bridges, engines on airplanes, your satellite receiver, weather sensors, your work ID card, and a whole lot more, all generate data.

In Chapter 2, we outline some of the popular problem domains and deployment patterns that suit Big Data technologies. We can't possibly cover all of the potential usage patterns, but we'll share experiences we've seen and hinted at earlier in this section. You'll find a recurring theme to Big Data opportunities—more data and data not easily analyzed before. In addition we will contrast and compare Big Data solutions with traditional warehouse solutions that are part of every IT shop. We will say it here and often within the book: Big Data complements existing analysis systems, it does not replace them (in this chapter we'll give you a good analogy that should get the point across quite vividly).

Without getting into the technology aspects, Chapter 3 talks about why we think IBM's Big Data platform is the best solution out there (yes, we work for IBM, but read the chapter; it's compelling!). If you take a moment to consider Big Data, you'll realize that it's not just about getting up and running with Hadoop (the key open source technology that provides a Big Data engine) and operationally managing it with a toolset. Consider this: we can't think of a single customer who gets excited about buying, managing, and installing technology. Our clients get excited about the opportunities their technologies allow them to exploit to their benefits; our customers have a vision of the picture they want to paint and we're going to help you turn into Claude Monet. IBM not only helps you flatten the time it takes to get Big Data up and running, but the fact that IBM has an offering in this space means it brings a whole lot more to the table: a platform. For example, if there's one concept that IBM is synonymous with, it is *enterprise class*. IBM understands fault tolerance, high availability, security, governance, and robustness. So when you step back from the open source Big Data Hadoop offering, you'll see that IBM is uniquely positioned to harden it for the enterprise. But BigInsights does more than just make Hadoop enterprise reliable and scalable; it makes the data stored in Hadoop easily exploitable without armies of Java programmers and Ph.D. statisticians. Consider that BigInsights adds analytic toolkits, resource management, compression, security, and more; you'll actually be able to take an enterprise-hardened Hadoop platform and quickly build a solution without having to buy piece parts or build the stuff yourself.

If you recall earlier in this foreword, we talked about how Big Data technologies are not a replacement for your current technologies—rather, they are a complement. This implies the obvious: you are going to have to *integrate* Big Data with the rest of your enterprise infrastructure, and you'll have *governance* requirements as well. What company understands data integration and governance better than IBM? It's a global economy, so if you think *language nationalization*, IBM should come to mind. (Is a text analytics platform only for English-based analysis? We hope not!) Think Nobel-winning world-class researchers, mathematicians, statisticians, and more: there's lots of this caliber talent in the halls of IBM, many working on Big Data problems. Think Watson (famous for its winning *Jeopardy!* performance) as a proof point of what IBM is capable of providing. Of course, you're going to want support for your Big Data platform, and who can provide *direct-to-engineer support, around the world, in a 24×7 manner?* What are you going to do with your Big Data? Analyze it! The *lineage of IBM's data analysis platforms* (SPSS, Cognos, Smart Analytics Systems, Netezza, text annotators, speech-to-text, and so much more—IBM has spent over $14 billion in the last five years on analytic acquisitions alone) offer immense opportunity for year-after-year extensions to its Big Data platform.

Of course we would be remiss not to mention how dedicated IBM is to the open source community in general. IBM has a rich heritage of supporting open source. Contributions such as the de facto standard integrated development environment (IDE) used in open source—Eclipse, Unstructured Information Management Architecture (UIMA), Apache Derby, Lucene, XQuery, SQL, and Xerces XML processor—are but a few of the too many to mention. We want to make one thing very clear—IBM is committed to Hadoop open source. In fact, Jaql (you will learn about this in Chapter 4) was donated to the open source Hadoop community by IBM. Moreover, IBM is continually working on additional technologies for potential Hadoop-related donations. Our development labs have Hadoop committers that work alongside other Hadoop committers from Facebook, LinkedIn, and more. Finally, you are likely to find one of our developers on any Hadoop forum. We believe IBM's commitment to open source Hadoop, combined with its vast intellectual property and research around enterprise needs and analytics, delivers a true Big Data platform.

Part II—Big Data: From the Technology Perspective starts by giving you some basics about Big Data open source technologies in Chapter 4. This chapter lays the "ground floor" with respect to open source technologies that are synonymous with Big Data—the most common being *Hadoop* (an Apache top-level project whose execution engine is behind the Big Data movement). You're not going to be a Hadoop expert after reading this chapter, but you're going to have a basis for understanding such terms as *Pig, Hive, HDFS, MapReduce,* and *ZooKeeper,* among others.

Chapter 5 is one of the most important chapters in this book. This chapter introduces you to the concept that splits Big Data into two key areas that only IBM seems to be talking about when defining Big Data: *Big Data in motion* and *Big Data at rest*. In this chapter, we focus on the at-rest side of the Big Data equation and IBM's InfoSphere BigInsights (BigInsights), which is the enterprise capable Hadoop platform from IBM. We talk about the IBM technologies we alluded to in Chapter 3—only with technical explanations and illustrations into how IBM differentiates itself with its Big Data platform. You'll learn about how IBM's General Parallel File system (GPFS), synonymous with enterprise class, has been extended to participate in a Hadoop environment as GPFS shared nothing cluster (SNC). You'll learn about how IBM's BigInsights platform includes a text analytics toolkit with a rich annotation development environment that lets you build or customize text annotators without having to use Java or some other programming language. You'll learn about fast data compression without GPL licensing concerns in the Hadoop world, special high-speed database connector technologies, machine learning analytics, management tooling, a flexible workload governor that provides a richer business policy–oriented management framework than the default Hadoop workload manager, security lockdown, enhancing MapReduce with intelligent adaptation, and more. After reading this chapter, we think the questions or capabilities you will want your Big Data provider to answer will change and will lead you to ask questions that prove your vendor actually has a real Big Data platform. We truly believe your Big Data journey needs to start with a Big Data platform—powerful analytics tooling that sits on top of world class enterprise-hardened and capable technology.

In Chapter 6 we finish off the book by covering the other side of the Big Data "coin": analytics on data in motion. Chapter 6 introduces you to IBM InfoSphere Streams (Streams), in some depth, along with examples from real clients and how they are using Streams to realize better business outcomes, make better predictions, gain a competitive advantage for their company, and even improve the health of our most fragile. We also detail how Streams works, a special streams processing language built to flatten the time it takes to write Streams applications, how it is configured, and the components of a stream (namely operators and adapters). In much the same way as BigInsights makes Hadoop enterprise-ready, we round off the chapter detailing the capabilities that make Streams enterprise-ready, such as high availability, scalability, ease of use, and how it integrates into your existing infrastructure.

We understand that you will spend the better part of a couple of hours of your precious time to read this book; we're confident by the time you are finished, you'll have a good handle on the Big Data opportunity that lies ahead, a better understanding of the requirements that will ensure that you have the right Big Data platform, and a strong foundational knowledge as to the business opportunities that lie ahead with Big Data and some of the technologies available.

When we wrote this book, we had to make some tough trade-offs because of its limited size. These decisions were not easy; sometimes we felt we were cheating the technical reader to help the business reader, and sometimes we felt the opposite. In the end, we hope to offer you a fast path to Big Data knowledge and understanding of the unique position IBM is in to make it more and more of a reality in your place of business.

As you travel the roads of your Big Data journey, we think you will find something that you didn't quite expect when you first started it; since it's not an epic movie, we'll tell you now and in a year from now, let us know if we were right. We think you'll find that not only will Big Data technologies become a rich repository commonplace in the enterprise, *but* also an application platform (akin to WebSphere). You'll find the need for declarative languages that can be used to build analytic applications in a rich ecosystem that is more integrated than ever into where the data is stored. You'll find yourself in need of object classes that provide specific kinds of analytics and you'll demand a development environment that lets you reuse components and customize at will. You'll require methods to deploy these applications (in a concept similar to Blackberry's AppWorld or Apple's AppStore), visualization capabilities, and more.

As you can see, this book isn't too big (it was never meant to be a novel), and it's got five authors. When we first met, one of us quipped that the first thing that came to his mind was how writing this book was perhaps like a customer visit: lots of IBMers at the table. But you know what? That's the power of this company: its ability to reach across experiences that span billions of dollars of transactions, across varying industries, and broad expertise. Our authoring team has more than 100 years of collective experience and many thousands of hours of consulting and customer interactions. We've had experiences in research, patents, competitive, management, development, and various industry verticals. We hope that our group effectively shared some of that experience with you in this book as a start to your Big Data journey.

Part I

Big Data:
From the Business
Perspective

1

What Is Big Data? Hint: You're a Part of It Every Day

Where should we start a book on Big Data? How about with a definition, because the term "Big Data" is a bit of a misnomer since it implies that pre-existing data is somehow small (it isn't) or that the only challenge is its sheer size (size is one of them, but there are often more). In short, the term Big Data applies to information that can't be processed or analyzed using traditional processes or tools. Increasingly, organizations today are facing more and more Big Data challenges. They have access to a wealth of information, but they don't know how to get value out of it because it is sitting in its most raw form or in a semistructured or unstructured format; and as a result, they don't even know whether it's worth keeping (or even able to keep it for that matter). An IBM survey found that over half of the business leaders today realize they don't have access to the insights they need to do their jobs. Companies are facing these challenges in a climate where they have the ability to store anything and they are generating data like never before in history; combined, this presents a real information challenge. It's a conundrum: today's business has more access to potential insight than ever before, yet as this potential gold mine of data piles up, the percentage of data the business can process is going down—fast. We feel that before we can talk about all the great things you can do with Big Data, and how IBM has a unique end-to-end platform that we believe will make you more successful, we need to talk about the characteristics of Big Data and how it fits into the current information management landscape.

Quite simply, the Big Data era is in full force today because the world is changing. Through *instrumentation*, we're able to sense more things, and if we can sense it, we tend to try and store it (or at least some of it). Through advances in communications technology, people and things are becoming increasingly *interconnected*—and not just some of the time, but all of the time. This interconnectivity rate is a runaway train. Generally referred to as *machine-to-machine (M2M)*, interconnectivity is responsible for double-digit year over year (YoY) data growth rates. Finally, because small integrated circuits are now so inexpensive, we're able to add *intelligence* to almost everything.

Even something as mundane as a railway car has hundreds of sensors. On a railway car, these sensors track such things as the conditions experienced by the rail car, the state of individual parts, and GPS-based data for shipment tracking and logistics. After train derailments that claimed extensive losses of life, governments introduced regulations that this kind of data be stored and analyzed to prevent future disasters. Rail cars are also becoming more intelligent: processors have been added to interpret sensor data on parts prone to wear, such as bearings, to identify parts that need repair before they fail and cause further damage—or worse, disaster. But it's not just the rail cars that are intelligent—the actual rails have sensors every few feet. What's more, the data storage requirements are for the whole ecosystem: cars, rails, railroad crossing sensors, weather patterns that cause rail movements, and so on. Now add this to tracking a rail car's cargo load, arrival and departure times, and you can very quickly see you've got a Big Data problem on your hands. Even if every bit of this data was relational (and it's not), it is all going to be raw and have very different formats, which makes processing it in a traditional relational system impractical or impossible. Rail cars are just one example, but everywhere we look, we see domains with velocity, volume, and variety combining to create the Big Data problem.

IBM has created a whole model around helping businesses embrace this change via its Smart Planet platform. It's a different way of thinking that truly recognizes that the world is now instrumented, interconnected, and intelligent. The Smart Planet technology and techniques promote the understanding and harvesting of the world's data reality to provide opportunities for unprecedented insight and the opportunity to change the way things are done. To build a Smart Planet it's critical to harvest all the data, and the IBM Big Data platform is designed to do just that; in fact, it is a key architectural pillar of the Smart Planet initiative.

Characteristics of Big Data

Three characteristics define Big Data: *volume, variety*, and *velocity* (as shown in Figure 1-1). Together, these characteristics define what we at IBM refer to as "Big Data." They have created the need for a new class of capabilities to augment the way things are done today to provide better line of site and controls over our existing knowledge domains and the ability to act on them.

The IBM Big Data platform gives you the unique opportunity to extract insight from an immense volume, variety, and velocity of data, in context, beyond what was previously possible. Let's spend some time explicitly defining these terms.

Can There Be Enough? The Volume of Data

The sheer *volume* of data being stored today is exploding. In the year 2000, 800,000 petabytes (PB) of data were stored in the world. Of course, a lot of the data that's being created today isn't analyzed at all and that's another problem we're trying to address with BigInsights. We expect this number to reach 35 zettabytes (ZB) by 2020. Twitter alone generates more than 7 terabytes (TB) of data every day, Facebook 10 TB, and some enterprises generate

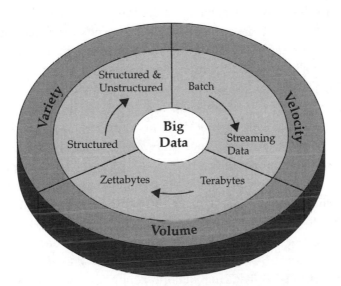

Figure 1-1 *IBM characterizes Big Data by its volume, velocity, and variety—or simply, V³.*

terabytes of data every hour of every day of the year. It's no longer unheard of for individual enterprises to have storage clusters holding petabytes of data. We're going to stop right there with the factoids: Truth is, these estimates will be out of date by the time you read this book, and they'll be further out of date by the time you bestow your great knowledge of data growth rates on your friends and families when you're done reading this book.

When you stop and think about it, it's little wonder we're drowning in data. If we can track and record something, we typically do. (And notice we didn't mention the analysis of this stored data, which is going to become a theme of Big Data—the newfound utilization of data we track and don't use for decision making.) We store everything: environmental data, financial data, medical data, surveillance data, and the list goes on and on. For example, taking your smartphone out of your holster generates an event; when your commuter train's door opens for boarding, that's an event; check in for a plane, badge into work, buy a song on iTunes, change the TV channel, take an electronic toll route—everyone of these actions generates data. Need more? The St. Anthony Falls Bridge (which replaced the 2007 collapse of the I-35W Mississippi River Bridge) in Minneapolis has more than 200 embedded sensors positioned at strategic points to provide a fully comprehensive monitoring system where all sorts of detailed data is collected and even a shift in temperature and the bridge's concrete reaction to that change is available for analysis. Okay, you get the point: There's more data than ever before and all you have to do is look at the terabyte penetration rate for personal home computers as the telltale sign. We used to keep a list of all the data warehouses we knew that surpassed a terabyte almost a decade ago—suffice to say, things have changed when it comes to volume.

As implied by the term "Big Data," organizations are facing massive volumes of data. Organizations that don't know how to manage this data are overwhelmed by it. But the opportunity exists, with the right technology platform, to analyze almost all of the data (or at least more of it by identifying the data that's useful to you) to gain a better understanding of your business, your customers, and the marketplace. And this leads to the current conundrum facing today's businesses across all industries. As the amount of data available to the enterprise is on the rise, the percent of data it can process, understand, and analyze is on the decline, thereby creating the blind zone you see in Figure 1-2. What's in that blind zone? You don't know: it might be

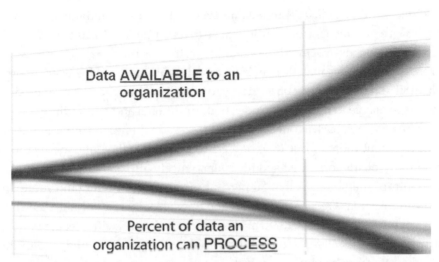

Figure 1-2   *The volume of data available to organizations today is on the rise, while the percent of data they can analyze is on the decline.*

something great, or may be nothing at all, but the "don't know" is the problem (or the opportunity, depending how you look at it).

The conversation about data volumes has changed from terabytes to petabytes with an inevitable shift to zettabytes, and all this data can't be stored in your traditional systems for reasons that we'll discuss in this chapter and others.

Variety Is the Spice of Life

The volume associated with the Big Data phenomena brings along new challenges for data centers trying to deal with it: its *variety*. With the explosion of sensors, and smart devices, as well as social collaboration technologies, data in an enterprise has become complex, because it includes not only traditional relational data, but also raw, semistructured, and unstructured data from web pages, web log files (including click-stream data), search indexes, social media forums, e-mail, documents, sensor data from active and passive systems, and so on. What's more, traditional systems can struggle to store and perform the required analytics to gain understanding from the contents of these logs because much of the information being generated doesn't lend itself to traditional database technologies. In our experience, although some companies are moving down the path, by and large, most are just beginning to understand the opportunities of Big Data (and what's at stake if it's not considered).

Quite simply, *variety* represents all types of data—a fundamental shift in analysis requirements from traditional structured data to include raw, semistructured, and unstructured data as part of the decision-making and insight process. Traditional analytic platforms can't handle variety. However, an organization's success will rely on its ability to draw insights from the various kinds of data available to it, which includes both traditional and nontraditional.

When we look back at our database careers, sometimes it's humbling to see that we spent more of our time on just 20 percent of the data: the relational kind that's neatly formatted and fits ever so nicely into our strict schemas. But the truth of the matter is that 80 percent of the world's data (and more and more of this data is responsible for setting new velocity and volume records) is unstructured, or semistructured at best. If you look at a Twitter feed, you'll see structure in its JSON format—but the actual text is not structured, and understanding that can be rewarding. Video and picture images aren't easily or efficiently stored in a relational database, certain event information can dynamically change (such as weather patterns), which isn't well suited for strict schemas, and more. To capitalize on the Big Data opportunity, enterprises must be able to analyze *all* types of data, both relational and nonrelational: text, sensor data, audio, video, transactional, and more.

How Fast Is Fast? The Velocity of Data

Just as the sheer volume and variety of data we collect and store has changed, so, too, has the *velocity* at which it is generated and needs to be handled. A conventional understanding of velocity typically considers how quickly the data is arriving and stored, and its associated rates of retrieval. While managing all of that quickly is good—and the volumes of data that we are looking at are a consequence of how quick the data arrives—we believe the idea of velocity is actually something far more compelling than these conventional definitions.

To accommodate velocity, a new way of thinking about a problem must start at the inception point of the data. Rather than confining the idea of velocity to the growth rates associated with your data repositories, we suggest you apply this definition to data in motion: The speed at which the data is flowing. After all, we're in agreement that today's enterprises are dealing with petabytes of data instead of terabytes, and the increase in RFID sensors and other information streams has led to a constant flow of data at a pace that has made it impossible for traditional systems to handle.

Sometimes, getting an edge over your competition can mean identifying a trend, problem, or opportunity only seconds, or even microseconds, before someone else. In addition, more and more of the data being produced today has a very short shelf-life, so organizations must be able to analyze this data in near real time if they hope to find insights in this data. Big Data scale *streams computing* is a concept that IBM has been delivering on for some time and serves as a new paradigm for the Big Data problem. In traditional processing, you can think of running queries against relatively static data: for example, the query "Show me all people living in the New Jersey flood zone" would result in a single result set to be used as a warning list of an incoming weather pattern. With streams computing, you can execute a process similar to a continuous query that identifies people who are *currently* "in the New Jersey flood zones," but you get continuously updated results, because location information from GPS data is refreshed in real time.

Dealing effectively with Big Data requires that you perform analytics against the volume and variety of data while it is *still in motion*, not just after it is *at rest*. Consider examples from tracking neonatal health to financial markets; in every case, they require handling the volume and variety of data in new ways. The velocity characteristic of Big Data is one key differentiator that makes IBM the best choice for your Big Data platform. We define it as an *inclusional* shift from solely batch insight (Hadoop style) to batch insight combined with streaming-on-the-wire insight, and IBM seems to be the only vendor talking about velocity being more than how fast data is generated (which is really part of the volume characteristic).

Now imagine a cohesive Big Data platform that can leverage the best of both worlds and take streaming real-time insight to spawn further research based on emerging data. As you think about this, we're sure you'll start to share the same excitement we have around the unique proposition available with an IBM Big Data platform.

Data in the Warehouse and Data in Hadoop (It's Not a Versus Thing)

In our experience, traditional warehouses are mostly ideal for analyzing structured data from various systems and producing insights with known and relatively stable measurements. On the other hand, we feel a Hadoop-based platform is well suited to deal with semistructured and unstructured data, as well as when a data discovery process is needed. That isn't to say that Hadoop can't be used for structured data that is readily available in a raw format; because it can, and we talk about that in Chapter 2.

In addition, when you consider *where* data should be stored, you need to understand *how* data is stored today and what features characterize your persistence options. Consider your experience with storing data in a traditional data warehouse. Typically, this data goes through a lot of rigor to make it into the warehouse. Builders and consumers of warehouses have it etched in their minds that the data they are looking at in their warehouses must shine with respect to quality; subsequently, it's cleaned up via cleansing, enrichment, matching, glossary, metadata, master data management, modeling, and other services before it's ready for analysis. Obviously, this can be an expensive process. Because of that expense, it's clear that the data that lands in the warehouse is deemed not just of high value, but it has a broad purpose: it's going to go places and will be used in reports and dashboards where the accuracy of that data is key. For example, Sarbanes-Oxley (SOX) compliance, introduced in 2002, requires the CEO and CFO of publicly traded companies on U.S.-based exchanges to certify the accuracy of their financial statements (Section 302, "Corporate Responsibility for Financial Reports"). There are serious (we're talking the potential for jail time here) penalties associated if the data being reported isn't accurate or "true." Do you think these folks are going to look at reports of data that aren't pristine?

In contrast, Big Data repositories rarely undergo (at least initially) the full quality control rigors of data being injected into a warehouse, because not only is prepping data for some of the newer analytic methods characterized by Hadoop use cases cost prohibitive (which we talk about in the next chapter), but the data isn't likely to be distributed like data warehouse data. We could say that data warehouse data is trusted enough to be "public," while Hadoop data isn't as trusted (*public* can mean vastly distributed within the company and not for external consumption), and although this will likely change in the future, today this is something that experience suggests characterizes these repositories.

Our experiences also suggest that in today's IT landscape, specific pieces of data have been stored based on their perceived value, and therefore any information beyond those preselected pieces is unavailable. This is in contrast to a

Hadoop-based repository scheme where the entire business entity is likely to be stored and the fidelity of the Tweet, transaction, Facebook post, and more is kept intact. Data in Hadoop might seem of low value today, or its value nonquantified, but it can in fact be the key to questions yet unasked. IT departments pick and choose high-valued data and put it through rigorous cleansing and transformation processes because they know that data has a *high known value per byte* (a relative phrase, of course). Why else would a company put that data through so many quality control processes? Of course, since the value per byte is high, the business is willing to store it on relatively higher cost infrastructure to enable that interactive, often public, navigation with the end user communities, and the CIO is willing to invest in cleansing the data to increase its value per byte.

With Big Data, you should consider looking at this problem from the opposite view: With all the volume and velocity of today's data, there's just no way that you can afford to spend the time and resources required to cleanse and document every piece of data properly, because it's just not going to be economical. What's more, how do you know if this Big Data is even valuable? Are you going to go to your CIO and ask her to increase her capital expenditure (CAPEX) and operational expenditure (OPEX) costs by fourfold to quadruple the size of your warehouse on a hunch? For this reason, we like to characterize the initial nonanalyzed raw Big Data as having a *low value per byte*, and, therefore, until it's proven otherwise, you can't afford to take the path to the warehouse; however, given the vast amount of data, the potential for great insight (and therefore greater competitive advantage in your own market) is quite high if you can analyze all of that data.

At this point, it's pertinent to introduce the idea of *cost per compute*, which follows the same pattern as the value per byte ratio. If you consider the focus on the quality data in traditional systems we outlined earlier, you can conclude that the cost per compute in a traditional data warehouse is relatively high (which is fine, because it's a proven and known higher value per byte), versus the cost of Hadoop, which is low.

Of course, other factors can indicate that certain data might be of high value yet never make its way into the warehouse, or there's a desire for it to make its way out of the warehouse into a lower cost platform; either way, you might need to cleanse some of that data in Hadoop, and IBM can do that (a key differentiator). For example, unstructured data can't be easily stored in a warehouse.

Indeed, some warehouses are built with a predefined corpus of questions in mind. Although such a warehouse provides some degree of freedom for query and mining, it could be that it's constrained by what is in the schema (most unstructured data isn't found here) and often by a performance envelope that can be a functional/operational hard limit. Again, as we'll reiterate often in this book, we are not saying a Hadoop platform such as IBM InfoSphere BigInsights is a replacement for your warehouse; instead, it's a complement.

A Big Data platform lets you store all of the data in its native business object format and get value out of it through massive parallelism on readily available components. For your interactive navigational needs, you'll continue to pick and choose sources and cleanse that data and keep it in warehouses. But you can get more value out of analyzing more data (that may even initially seem unrelated) in order to paint a more robust picture of the issue at hand. Indeed, data might sit in Hadoop for a while, and when you discover its value, it might migrate its way into the warehouse when its value is proven and sustainable.

Wrapping It Up

We'll conclude this chapter with a gold mining analogy to articulate the points from the previous section and the Big Data opportunity that lies before you. In the "olden days" (which, for some reason, our kids think is a time when we were their age), miners could actually *see* nuggets or veins of gold; they clearly appreciated the value and would dig and sift near previous gold finds hoping to strike it rich. That said, although there was more gold out there—it could have been in the hill next to them or miles away—it just wasn't visible to the naked eye, and it became a gambling game. You dug like crazy near where gold was found, but you had no idea whether more gold would be found. And although history has its stories of gold rush fevers, nobody mobilized millions of people to dig everywhere and anywhere.

In contrast, today's gold rush works quite differently. Gold mining is executed with massive capital equipment that can process millions of tons of dirt that is worth nothing. Ore grades of 30 mg/kg (30 ppm) are usually needed before gold is visible to the naked eye—that is, most gold in gold mines today is invisible. Although there is all this gold (high-valued data) in all this dirt (low-valued data), by using the right equipment, you can economically process lots of dirt and keep the flakes of gold you find. The flakes of gold are then taken for integration and put together to make a bar of gold, which is stored and logged in a place that's safe, governed, valued, and trusted.

This really is what Big Data is about. You can't afford to sift through all the data that's available to you in your traditional processes; it's just too much data with too little known value and too much of a gambled cost. The IBM Big Data platform gives you a way to economically store and process all that data and find out what's valuable and worth exploiting. What's more, since we talk about analytics for data at rest and data in motion, the actual data from which you can find value is not only broader with the IBM Big Data platform, but you're able to use and analyze it more quickly in real time.

2

Why Is Big Data Important?

This chapter's title describes exactly what we're going to cover: Why is Big Data important? We're also going to discuss some of our real customer experiences, explaining how we've engaged and helped develop new applications and potential approaches to solving previously difficult—if not impossible—challenges for our clients. Finally, we'll highlight a couple usage patterns that we repeatedly encounter in our engagements that cry out for the kind of help IBM's Big Data platform, comprised of IBM InfoSphere BigInsights (BigInsights) and IBM InfoSphere Streams (Streams), can offer.

When to Consider a Big Data Solution

The term *Big Data* can be interpreted in many different ways and that's why in Chapter 1 we defined Big Data as conforming to the volume, velocity, and variety (V^3) attributes that characterize it. Note that Big Data solutions aren't a replacement for your existing warehouse solutions, and in our humble opinion, any vendor suggesting otherwise likely doesn't have the full gambit of experience or understanding of your investments in the traditional side of information management.

We think it's best to start out this section with a couple of key Big Data principles we want you to keep in mind, before outlining some considerations as to when you use Big Data technologies, namely:

- Big Data solutions are ideal for analyzing not only raw structured data, but semistructured and unstructured data from a wide variety of sources.

- Big Data solutions are ideal when all, or most, of the data needs to be analyzed versus a sample of the data; or a sampling of data isn't nearly as effective as a larger set of data from which to derive analysis.

- Big Data solutions are ideal for iterative and exploratory analysis when business measures on data are not predetermined.

When it comes to solving information management challenges using Big Data technologies, we suggest you consider the following:

- Is the reciprocal of the traditional analysis paradigm appropriate for the business task at hand? Better yet, can you see a Big Data platform complementing what you currently have in place for analysis and achieving synergy with existing solutions for better business outcomes?

  For example, typically, data bound for the analytic warehouse has to be cleansed, documented, and trusted before it's neatly placed into a strict warehouse schema (and, of course, if it can't fit into a traditional row and column format, it can't even get to the warehouse in most cases). In contrast, a Big Data solution is not only going to leverage data not typically suitable for a traditional warehouse environment, and in massive amounts of volume, but it's going to give up some of the formalities and "strictness" of the data. The benefit is that you can preserve the fidelity of data and gain access to mountains of information for exploration and discovery of business insights *before* running it through the due diligence that you're accustomed to; the data that can be included as a participant of a cyclic system, enriching the models in the warehouse.

- Big Data is well suited for solving information challenges that don't natively fit within a traditional relational database approach for handling the problem at hand.

It's important that you understand that conventional database technologies *are* an important, and relevant, part of an overall analytic solution. In fact, they become even more vital when used in conjunction with your Big Data platform.

A good analogy here is your left and right hands; each offers individual strengths and optimizations for a task at hand. For example, if you've ever played baseball, you know that one hand is better at throwing and the other at catching. It's likely the case that each hand could try to do the other task that it isn't a natural fit for, but it's very awkward (try it; better yet, film yourself trying it and you will see what we mean). What's more, you don't see baseball players catching with one hand, stopping, taking off their gloves, and throwing with the same hand either. The left and right hands of a baseball player work in unison to deliver the best results. This is a loose analogy to traditional database and Big Data technologies: Your information platform shouldn't go into the future without these two important entities working together, because the outcomes of a cohesive analytic ecosystem deliver premium results in the same way your coordinated hands do for baseball. There exists some class of problems that don't natively belong in traditional databases, at least not at first. And there's data that we're not sure we want in the warehouse, because perhaps we don't know if it's rich in value, it's unstructured, or it's too voluminous. In many cases, we can't find out the value per byte of the data until *after* we spend the effort and money to put it into the warehouse; but we want to be sure that data is worth saving and has a high value per byte before investing in it.

Big Data Use Cases: Patterns for Big Data Deployment

This chapter is about helping you understand why Big Data is important. We could cite lots of press references around Big Data, upstarts, and chatter, but that makes it sound more like a marketing sheet than an inflection point. We believe the best way to frame why Big Data is important is to share with you a number of our real customer experiences regarding usage patterns they are facing (and problems they are solving) with an IBM Big Data platform. These patterns represent great Big Data opportunities—business problems that weren't easy to solve before—and help you gain an understanding of how Big Data can help you (or how it's helping your competitors make you less competitive if you're not paying attention).

In our experience, the IBM BigInsights platform (which embraces Hadoop and extends it with a set of rich capabilities, which we talk about later in this book) is applicable to every industry we serve. We could cover hundreds of use cases in this chapter, but in the interest of space, we'll discuss six that expose some of the most common usage patterns we see. Although the explanations of the usage patterns might be industry-specific, many are broadly cross-industry applicable (which is how we settled on them). You'll find a common trait in all of the usage patterns discussed here: They all involve a new way of doing things that is now more practical and finally possible with Big Data technologies.

IT for IT Log Analytics

Log analytics is a common use case for an inaugural Big Data project. We like to refer to all those logs and trace data that are generated by the operation of your IT solutions as *data exhaust*. Enterprises have lots of data exhaust, and it's pretty much a pollutant if it's just left around for a couple of hours or days in case of emergency and simply purged. Why? Because we believe data exhaust has concentrated value, and IT shops need to figure out a way to store and extract value from it. Some of the value derived from data exhaust is obvious and has been transformed into value-added click-stream data that records every gesture, click, and movement made on a web site.

Some data exhaust value isn't so obvious. At the DB2 development labs in Toronto (Ontario, Canada) engineers derive terrific value by using BigInsights for performance optimization analysis. For example, consider a large, clustered transaction-based database system and try to preemptively find out where small optimizations in correlated activities across separate servers might be possible. There are needles (some performance optimizations) within a haystack (mountains of stack trace logs across many servers). Trying to find correlation across tens of gigabytes of per core stack trace information is indeed a daunting task, but a Big Data platform made it possible to identify previously unreported areas for performance optimization tuning.

Quite simply, IT departments need logs at their disposal, and today they just can't store enough logs and analyze them in a cost-efficient manner, so logs are typically kept for emergencies and discarded as soon as possible. Another reason why IT departments keep large amounts of data in logs is to look for rare problems. It is often the case that the most common problems are known and easy to deal with, but the problem that happens "once in a while" is typically more difficult to diagnose and prevent from occurring again.

We think that IT yearns (or should yearn) for log longevity. We also think that both business and IT know there is value in these logs, and that's why we often see lines of business duplicating these logs and ending up with scattershot retention and nonstandard (or duplicative) analytic systems that vary greatly by team. Not only is this ultimately expensive (more aggregate data needs to be stored—often in expensive systems), but since only slices of the data are available, it is nearly impossible to determine holistic trends and issues that span such a limited retention time period and views of the information.

Today this log history can be retained, but in most cases, only for several days or weeks at a time, because it is simply too much data for conventional systems to store, and that, of course, makes it impossible to determine trends and issues that span such a limited retention time period. But there are more reasons why log analysis is a Big Data problem aside from its voluminous nature. The nature of these logs is semistructured and raw, so they aren't always suited for traditional database processing. In addition, log formats are constantly changing due to hardware and software upgrades, so they can't be tied to strict inflexible analysis paradigms. Finally, not only do you need to perform analysis on the longevity of the logs to determine trends and patterns and to pinpoint failures, but you need to ensure the analysis is done on all the data.

Log analytics is actually a pattern that IBM established after working with a number of companies, including some large financial services sector (FSS) companies. We've seen this use case come up with quite a few customers since; for that reason, we'll call this pattern *IT for IT*. If you can relate, we don't have to say anything else. If you're new to this usage pattern and wondering just who's interested in IT for IT Big Data solutions, you should know that this is an internal use case within an organization itself. For example, often non-IT business entities want this data provided to them as a kind of service bureau. An internal IT for IT implementation is well suited for any organization with a large data center footprint, especially if it is relatively complex. For example, service-oriented architecture (SOA) applications with lots of moving parts, federated data centers, and so on, all suffer from the same issues outlined in this section.

Customers are trying to gain better insights into how their systems are running and when and how things break down. For example, one financial firm we worked with affectionately refers to the traditional way of figuring out how an application went sideways as "Whac-A-Mole." When things go wrong in their heavily SOA-based environment, it's hard to determine what happened, because twenty-plus systems are involved in the processing of a certain transaction, making it really hard for the IT department to track down exactly why and where things went wrong. (We've all seen this movie: Everyone runs around the war room saying, "I didn't do it!"—there's also a scene in that movie where everyone is pointing their fingers at you.) We helped this client leverage a Big Data platform to analyze approximately 1TB of log data each day, with less than 5 minutes latency. Today, the client is able to decipher exactly what is happening across the entire stack with each and every transaction. When one of their customer's transactions, spawned from their mobile or Internet banking platforms goes wrong, they are able to tell exactly where and what component contributed to the problem. Of course, as you can imagine, this saves them a heck of a lot of time with problem resolution, without imposing additional monitoring inline with the transaction, because they are using the data exhaust that is already being generated as the source of analysis. But there's more to this use case than detecting problems: they are able to start to develop a base (or corpus) of knowledge so that they can better anticipate and understand the interaction between failures, their service bureau can generate best-practice remediation steps in the event of a specific problem, or better yet they can retune the infrastructure to eliminate them. This is about discoverable preventative maintenance, and that's potentially even more impactful.

Some of our large insurance and retail clients need to know the answers to such questions as, "What are the precursors to failures?", "How are these systems all related?", and more. You can start to see a cross-industry pattern here, can't you? These are the types of questions that conventional monitoring doesn't answer; a Big Data platform finally offers the opportunity to get some new and better insights into the problems at hand.

The Fraud Detection Pattern

Fraud detection comes up a lot in the financial services vertical, but if you look around, you'll find it in any sort of claims- or transaction-based environment (online auctions, insurance claims, underwriting entities, and so on). Pretty much anywhere some sort of financial transaction is involved presents a potential for misuse and the ubiquitous specter of fraud. If you leverage a Big Data platform, you have the opportunity to do more than you've ever done before to identify it or, better yet, stop it.

Several challenges in the fraud detection pattern are directly attributable to solely utilizing conventional technologies. The most common, and recurring, theme you will see across all Big Data patterns is limits on what can be stored as well as available compute resources to process your intentions. Without Big Data technologies, these factors limit what can be modeled. Less data equals constrained modeling. What's more, highly dynamic environments commonly have cyclical fraud patterns that come and go in hours, days, or weeks. If the data used to identify or bolster new fraud detection models isn't available with low latency, by the time you discover these new patterns, it's too late and some damage has already been done.

Traditionally, in fraud cases, samples and models are used to identify customers that characterize a certain kind of profile. The problem with this approach (and this is a trend that you're going to see in a lot of these use cases) is that although it works, you're profiling a segment and not the granularity at an individual transaction or person level. Quite simply, making a forecast based on a segment is good, but making a decision based upon the actual particulars of an individual transaction is obviously better. To do this, you need to work up a larger set of data than is conventionally possible in the traditional approach. In our customer experiences, we estimate that only 20 percent (or maybe less) of the available information that could be useful for fraud modeling is actually being used. The traditional approach is shown in Figure 2-1.

You're likely wondering, "Isn't the answer simply to load that other 80 percent of the data into your traditional analytic warehouse?" Go ahead and ask your CIO for the CAPEX and OPEX approvals to do this: you're going to realize quickly that it's too expensive of a proposition. You're likely thinking it will pay for itself with better fraud detection models, and although that's indeed the end goal, how can you be sure this newly loaded, cleansed, documented, and governed data was valuable in the first place (before the money is spent)? And therein is the point: You can use BigInsights to provide an elastic and cost-effective repository to establish what of the remaining 80 percent of the information is useful for fraud modeling, and then feed newly discovered high-value information back into the fraud model (it's the whole baseball left hand–right hand thing we referenced earlier in this chapter) as shown in Figure 2-2.

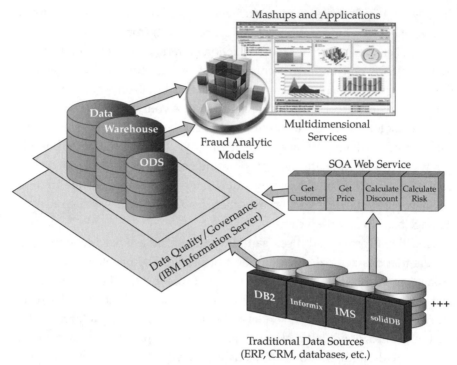

Figure 2-1   *Traditional fraud detection patterns use approximately 20 percent of available data.*

You can see in Figure 2-2 a modern-day fraud detection ecosystem that provides a low-cost Big Data platform for exploratory modeling and discovery. Notice how this data can be leveraged by traditional systems either directly or through integration into existing data quality and governance protocols. Notice the addition of InfoSphere Streams (the circle by the DB2 database cylinder) as well, which showcases the unique Big Data platform that only IBM can deliver: it's an ecosystem that provides analytics for data-in-motion and data-at-rest.

We teamed with a large credit card issuer to work on a permutation of Figure 2-2, and they quickly discovered that they could not only improve just how quickly they were able to speed up the build and refresh of their fraud detection models, but their models were broader and more accurate because of all the new insight. In the end, this customer took a process that once took about three weeks from when a transaction hit the transaction switch until when it was actually available for their fraud teams to work on, and turned

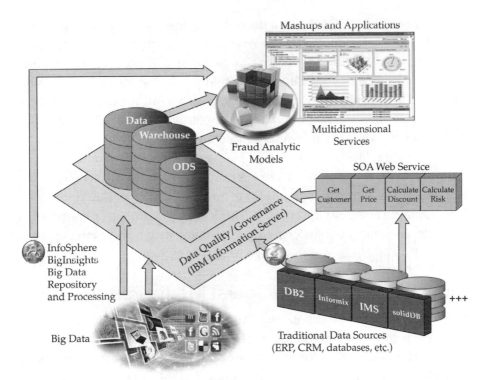

Figure 2-2   *A modern-day fraud detection ecosystem synergizes a Big Data platform with traditional processes.*

that latency into a couple of hours. In addition, the fraud detection models were built on an expanded amount of data that was roughly 50 percent broader than the previous set of data. As we can see in this example, all of that "80 percent of the data" that we talked about not being used wasn't all valuable in the end, but they found out what data had value and what didn't, in a cost-effective and efficient manner, using the BigInsights platform. Now, of course, once you have your fraud models built, you'll want to put them into action to try and prevent the fraud in the first place. Recovery rates for fraud are dismal in all industries, so it's best to prevent it versus discover it and try to recover the funds post-fraud. This is where InfoSphere Streams comes into play as you can see in Figure 2-2. Typically, fraud detection works after a transaction gets stored only to get pulled out of storage and analyzed; storing something to instantly pull it back out again feels like latency to us. With Streams, you can apply your fraud detection models as the transaction is happening.

In this section we focused on a financial services credit card company because it was an early one-on-one experience we had when first starting in Big Data. You shouldn't consider the use cases outlined in this section limited to what we've presented here; in fact, we told you at the start of this chapter that there are literally hundreds of usage patterns but we can't cover them all. In fact, fraud detection has massive applicability. Think about fraud in health care markets (health insurance fraud, drug fraud, medical fraud, and so on) and the ability to get in front of insurer and government fraud schemes (both claimants and providers). There's quite the opportunity there when the Federal Bureau of Investigation (FBI) estimates that health care fraud costs U.S. taxpayers over $60 billion a year. Think about fraudulent online product or ticket sales, money transfers, swiped banking cards, and more: you can see that the applicability of this usage pattern is extreme.

They Said What? The Social Media Pattern

Perhaps the most talked about Big Data usage pattern is social media and customer sentiment. You can use Big Data to figure out what customers are saying about you (and perhaps what they are saying about your competition); furthermore, you can use this newly found insight to figure out how this sentiment impacts the decisions you're making and the way your company engages. More specifically, you can determine how sentiment is impacting sales, the effectiveness or receptiveness of your marketing campaigns, the accuracy of your marketing mix (product, price, promotion, and placement), and so on.

Social media analytics is a pretty hot topic, so hot in fact that IBM has built a solution specifically to accelerate your use of it: Cognos Consumer Insights (CCI). It's a point solution that runs on BigInsights and it's quite good at what it does. CCI can tell you what people are saying, how topics are trending in social media, and all sorts of things that affect your business, all packed into a rich visualization engine.

Although basic insights into social media can tell you what people are saying and how sentiment is trending, they can't answer what is ultimately a more important question: "Why are people saying what they are saying and behaving in the way they are behaving?" Answering this type of question requires enriching the social media feeds with additional and differently shaped information that's likely residing in other enterprise systems. Simply put, linking behavior, and the driver of that behavior, requires relating social media analytics back to your traditional data repositories, whether they are SAP, DB2, Teradata, Oracle, or something else. You have to look beyond just the data; you have to look at the interaction of what people are doing with their behaviors, current financial trends, actual transactions that you're seeing internally, and so on. Sales, promotions, loyalty programs, the merchandising mix, competitor actions, and even variables such as the weather can all be drivers for what consumers feel and how opinions are formed. Getting to the core of why your customers are behaving a certain way requires merging information types in a dynamic and cost-effective way, especially during the initial exploration phases of the project.

Does it work? Here's a real-world case: A client introduced a different kind of environmentally friendly packaging for one of its staple brands. Customer sentiment was somewhat negative to the new packaging, and some months later, after tracking customer feedback and comments, the company discovered an unnerving amount of discontent around the change and moved to a different kind of eco-friendly package. It works, and we credit this progressive company for leveraging Big Data technologies to discover, understand, and react to the sentiment.

We'll hypothesize that if you don't have some kind of micro-blog oriented customer sentiment pulse-taking going on at your company, you're likely losing customers to another company that does.

NOTE   One of this book's authors is a prolific Facebook poster (for some reason he thinks the world is interested in his daily thoughts and experiences); after a number of consecutive flight delays that found their way to his Facebook wall, he was contacted by the airline to address the sentiment it detected. The airline acknowledged the issue; and although we won't get into what they did for him (think more legroom), the mere fact that they reached out to him meant someone was listening, which somehow made things better.

We think watching the world record Twitter tweets per second (Ttps) index is a telling indicator on the potential impact of customer sentiment. Super Bowl 2011 set a Twitter Ttps record in February 2011 with 4064 Ttps; it was surpassed by the announcement of bin Laden's death at 5106 Ttps, followed by the devastating Japan earthquake at 6939 Ttps. This Twitter record fell to the sentiment expressed when Paraguay's football penalty shootout win over Brazil in the Copa America quarterfinal peaked at 7166 Ttps, which could not beat yet another record set on the same day: a U.S. match win in the FIFA Women's World Cup at 7196 Ttps. When we went to print with this book, the famous singer Beyonce's Twitter announcement of her pregnancy peaked at 8868 Ttps and was the standing Ttps record. We think these records are very telling—not just because of the volume and velocity growth, but also because sentiment is being expressed for just about anything and everything, including your products and services. Truly, customer sentiment is everywhere; just ask Lady Gaga (@ladygaga) who is the most followed Tweeter in the world. What can we learn from this? First, everyone is able to express reaction and sentiment in seconds (often without thought or filters) for the world to see, and second, more and more people are expressing their thoughts or feelings about everything and anything.

The Call Center Mantra: "This Call May Be Recorded for Quality Assurance Purposes"

You're undoubtedly familiar with the title of this section: oddly enough, it seems that when we want our call with a customer service representative (CSR) to be recorded for quality assurance purposes, it seems the *may* part never works in our favor. The challenge of call center efficiencies is somewhat similar to the fraud detection pattern we discussed: Much like the fraud information latency critical to robust fraud models, if you've got experience in a call center, you'll know that the time/quality resolution metrics and trending discontent patterns for a call center can show up weeks after the fact. This latency means that if someone's on the phone and has a problem, you're not going to know about it right away from an enterprise perspective and you're not going to know that people are calling about this new topic or that you're seeing new and potentially disturbing trending in your interactions within a specific segment. The bottom line is this: In many cases, all of this call center information comes in too little, too late, and the problem is left solely up to the CSR to handle without consistent and approved remediation procedures in place.

We've been asked by a number of clients for help with this pattern, which we believe is well suited for Big Data. Call centers of all kinds want to find better ways to process information to address what's going on in the business with lower latency. This is a really interesting Big Data use case, because it uses analytics-in-motion and analytics-at-rest. Using in-motion analytics (Streams) means that you basically build your models and find out what's interesting based upon the conversations that have been converted from voice to text or with voice analysis as the call is happening. Using at-rest analytics (BigInsights), you build up these models and then promote them back into Streams to examine and analyze the calls that are actually happening in real time: it's truly a closed-loop feedback mechanism. For example, you use BigInsights to constantly run your analytics/heuristics/models against your data, and when new patterns are discovered, new business rules are created and pushed into the Streams model such that immediate action can be taken when a certain event occurs. Perhaps if a customer mentions a competitor, an alert is surfaced to the CSR to inform them of a current competitive promotion and a next best offer is generated for the client on the phone.

You can start to imagine all of the use cases that are at play here—capturing sentiment so that you know what people are saying, or expressing, or even volunteering information as to their disposition before your company takes a specific action; quite frankly, that's incredible insight. In addition, with more and more CSR outsourcing and a high probability that the CSR answering the phone isn't a native speaker of your language, nuances in discontent are not always easy to spot, and this kind of solution can help a call center improve its effectiveness.

Some industries or products have extremely disloyal customers with very high churn rates. For example, one of our clients competes in an industry characterized by a 70 percent annual customer churn rate (unlike North America, cell phone contracts aren't restrictive in other parts of the world). Even very small improvements in retention can be achieved by identifying which type of customers are most vulnerable, who they're calling, who's interested in a given topic, and so on; all of this has the potential to deliver a tremendous benefit to the business. With another customer, just a 2 percent difference in the conversion rates would double one of their offerings' revenue streams.

If you're able to capture and detect loyalty decay and work that into your CSR protocols, models, and canned remediation offers for a problem at hand, it can all lead to very happy outcomes in terms of either loss avoidance or additional revenue generation from satisfied customers open to potential cross-sell services. For example, once you've done the voice to text conversion of a call into BigInsights, you can then correlate that with everything from e-mails to social media and other things we've talked about in this chapter; you can even correlate it with back-office service quality reports to see if people are calling and expressing dissatisfaction with you based upon your back-end systems. If you are able to correlate and identify a pattern that shows where your systems have been slow or haven't behaved properly, and that just happened to be the reason why a particular individual is calling to cancel their services, but they never actually mention it, you can now find a correlation by what the customer is saying.

Here's a scenario we could all relate to: imagine calling a customer service department after getting disconnected twice and the agent saying, "We're sorry, we've been having some telephony issues and noticed you got disconnected twice…" How many times have you called in to complain about service with your high-speed Internet provider, and the CSR just dusted you off? The CSR didn't take action other than to listen. Does the service issue really get captured? Perhaps the agent handling the call fills out a form that provides a basic complaint of service, but does that get captured and correlated with point-in-time quality reports to indicate how the systems were running? Furthermore, we're working with customers to leverage the variety aspect of Big Data to correlate how trending in the call center is related to the rest of the business' operations. As an example, what types of calls and interactions are related to renewals, cross-sales, claims, and a variety of other key metrics in an insurance company? Few firms make those correlations today, but going forward they need to be able to do this to stay current with their competitors. How are you going to keep up, or, even better, lead in this area?

This is a really interesting Big Data use case, because it applies the art of the possible today using analytics in-motion and analytics at-rest, and is also a perfect fit for emerging capabilities like Watson. Using at-rest analytics (BigInsights) means that you basically build your models and find out what's interesting based upon the conversations that have been converted from voice to text or with voice analysis. Then you have the option of continuing to use at-rest analytics to harvest the call interactions in much lower latency (hours) compared to conventional operational cadence, or you build up these models and then promote them back into Streams to examine and analyze the calls as quickly as they can be converted to discover what is actually happening in near-real time. The results of the Streams analytics are flowed back into BigInsights—meaning it is truly a closed-loop feedback mechanism since BigInsights will then iterate over the results to improve the models. In the near future we see Watson being added into the mix to augment the pattern analytics that Streams is watching for to make expert recommendations on how to handle the interaction based on a much wider set of options than the call center agenda has available to them today.

As you can deduce from this pattern, a lot of "first-of-a-kind" capability potential for Big Data is present in a call center, and it's important that you start with some old-fashioned brainstorming. With the BigInsights platform the possibilities are truly limitless. Effectively analyzing once impossible to capture information is an established Big Data pattern that helps you understand things in a new way that ultimately relates back to what you're trying to do with your existing analytic systems.

Risk: Patterns for Modeling and Management

Risk modeling and management is another big opportunity and common Big Data usage pattern. Risk modeling brings into focus a recurring question when it comes to the Big Data usage patterns discussed in this chapter, "How much of your data do you use in your modeling?" The financial crisis of 2008, the associated subprime mortgage crisis, and its aftermath has made risk modeling and management a key area of focus for financial institutions. As you can tell by today's financial markets, a lack of understanding risk can have devastating wealth creation effects. In addition, newly legislated regulatory requirements affect financial institutions worldwide to ensure that their risk levels fall within acceptable thresholds.

As was the case in the fraud detection pattern, our customer engagements suggest that in this area, firms use between 15 and 20 percent of the available structured data in their risk modeling. It's not that they don't recognize that there's a lot of data that's potentially underutilized and rich in yet to be determined business rules that can be infused into a risk model; it's just that they don't know where the relevant information can be found in the rest of the data. In addition, as we've seen, it's just too expensive in many clients' current infrastructure to figure it out, because clearly they cannot double, triple, or quadruple the size of the warehouse just because there *might* (key word here) be some other information that's useful. What's more, some clients' systems just can't handle the increased load that up to 80 percent of the untapped data could bring, so even if they had the CAPEX and OPEX budgets to double or triple the size of the warehouse, many conventional systems couldn't handle the significant bursting of data and analytics that goes along with using the "rest of the data." Let's not forget that some data won't even fit into traditional systems, yet could be helpful in helping to model risk and you quickly realize you've got a conundrum that fundamentally requires a new approach.

Let's step back and think about what happens at the end of a trading day in a financial firm: They essentially get a closing snapshot of their positions. Using this snapshot, companies can derive insight and identify issues and concentrations using their models within a couple of hours and report back to regulators for internal risk control. For example, you don't want to find out something about your book of business in London that would impact trading in New York after the North American day's trading has begun. If you know about risks beforehand, you can do something about them as opposed to making the problem potentially worse because of what your New York bureau doesn't yet know or can't accurately predict. Now take this example and extend it to a broader set of worldwide financial markets (for example, add Asia into the mix), and you can see the same thing happens, except the risks and problems are compounded.

Two problems are associated with this usage pattern: "How much of the data will you use for your model?" (which we've already answered) and "How can you keep up with the data's velocity?" The answer to the second question, unfortunately, is often, "We can't." Finally, consider that financial services trend to move their risk model and dashboards to inter-day positions rather than just close-of-day positions, and you can see yet another challenge that can't be solved with traditional systems alone. Another characteristic of today's financial markets (other than us continually outward adjusting our planned retirement dates) is that there are massive trading volumes. If you mix massive spikes in volume, the requirements to better model and manage risk, and the inability to use all of the pertinent data in your models (let alone build them quickly or run them intra-day), you can see you've got a Big Data problem on your hands.

Big Data and the Energy Sector

The energy sector provides many Big Data use case challenges in how to deal with the massive volumes of sensor data from remote installations. Many companies are using only a fraction of the data being collected, because they lack the infrastructure to store or analyze the available scale of data.

Take for example a typical oil drilling platform that can have 20,000 to 40,000 sensors on board. All of these sensors are streaming data about the health of the oil rig, quality of operations, and so on. Not every sensor is actively broadcasting at all times, but some are reporting back many times per second. Now take a guess at what percentage of those sensors are actively utilized. If you're thinking in the 10 percent range (or even 5 percent), you're either a great guesser or you're getting the recurring theme for Big Data that spans industry and use cases: clients aren't using all of the data that's available to them in their decision-making process. Of course, when it comes to energy data (or any data for that matter) collection rates, it really begs the question, "If you've bothered to instrument the user or device or rig, in theory, you've done it on purpose, so why are you not capturing and leveraging the information you are collecting?"

With the thought of profit, safety, and efficiency in mind, businesses should be constantly looking for signals and be able to correlate those signals with their potential or probable outcomes. If you discard 90 percent of the sensor data as noise, you can't possibly understand or model those correlations. The "sensor noise bucket" is only as big as it is because of the lack of ability to store and analyze everything; folks here need a solution that allows for the separation of true signals from the noise. Of course, it's not enough to capture the data, be it noise or signals. You have to figure out the insight (and purge the noise), and the journey can't end there: you must be able to take action on this valuable insight. This is yet another great example of where data-in-motion analytics and data-at-rest analytics form a great Big Data left hand–right hand synergy: you have to take action on the identified valuable data while it's at rest (such as building models) and also take action while things are actually happening: a great data-in-motion Streams use case.

One BigInsights customer in Denmark, Vestas, is an energy sector global leader whose slogan decrees, "Wind. It means the world to us." Vestas is primarily engaged in the development, manufacturing, sale, and maintenance of power systems that use wind energy to generate electricity through its wind turbines. Its product range includes land and offshore wind turbines. At the time we wrote this book, it had more than 43,000 wind turbines in 65 countries on 5 continents. To us, it was great to get to know Vestas, because their vision is about the generation of clean energy, and they are using the IBM BigInsights platform as a method by which they can more profitably and efficiently generate even more clean energy, and that just makes us proud.

The alternative energy sector is very competitive and exploding in terms of demand. It also happens to be characterized by extreme competitive pricing, so any advantage you can get, you take in this market. Wind turbines, as it turns out, are multimillion-dollar investments with a lifespan of 20 to 30 years. That kind of caught us off guard. We didn't realize the effort that goes into their placement and the impact of getting a turbine placement wrong. The location chosen to install and operate a wind turbine can obviously greatly impact the amount of power generated by the unit, as well as how long it's able to remain in operation. To determine the optimal placement for a wind turbine, a large number of location-dependent factors must be considered, such as temperature, precipitation, wind velocity, humidity, atmospheric pressure, and more. This kind of data problem screams for a Big Data platform. Vestas's modeling system is expected to initially require 2.6 PB (2600 TB) of capacity, and as their engineers start developing their own forecasts and recording actual data of each wind turbine installation, their data capacity requirements are projected to increase to a whopping 6 PB (6000 TB)!

Vestas's legacy process for analyzing this data did not support the use of a full data set (there's that common theme when it comes to problems solved by a Big Data platform); what's more, it took them several weeks to execute the model. Vestas realized that they had a Big Data challenge that might be addressed by a Hadoop-based solution. The company was looking for a solution that would allow them to leverage all of the available data they collected to flatten the modeling time curve, support future expansion in modeling techniques, and improve the accuracy of decisions for wind turbine placement. After considering several other vendors, Vestas approached IBM for an enterprise-ready Hadoop-based Big Data analytics platform that embraces open source components and extends them the IBM enhancements outlined in Part II of this book (think a fully automated installation, enterprise hardening of

Hadoop, text and statistical-based analytics, governance, enterprise integration, development tooling, resource governance, visualization tools, and more: a platform).

Using InfoSphere BigInsights on IBM System x servers, Vestas is able to manage and analyze weather and location data in ways that were previously not possible. This allows them to gain insights that will lead to improved decisions for wind turbine placement and operations. All of this analysis comes from their Wind and Site Competency Center, where Vestas engineers are continually modeling weather data to forecast optimal turbine locations based on a global set of 1-by-1-kilometer grids (organized by country) that track and analyze *hundreds* of variables (temperature, barometric pressure, humidity, precipitation, wind direction, wind velocity at the ground level up to 300 feet, global deforestation metrics, satellite images, historical metrics, geospatial data, as well as data on phases of the moon and tides, and more). When you look at just a sample of the variables Vestas includes in their forecasting models, you can see the volume (PBs of data), velocity (all this data continually changes and streams into the data center as fast as the weather changes), and variety (all different formats, some structured, some unstructured—and most of it raw) that characterize this to be a Big Data problem solved by a partnership with IBM's Smarter Energy initiatives based on the IBM Big Data platform.

3

Why IBM for Big Data?

How many times have you heard, "This changes everything," only for history to show that, in fact, nothing much changed at all? We want to be clear on something here (and we'll repeat this important point throughout this book to ensure there is no doubt): Big Data technologies are important and we'll go so far as to call them a critical path for nearly all large organizations going forward, but traditional data platforms aren't going away—they're just getting a great partner.

Some historical context is useful here: A few years back, we heard that Hadoop would "change everything," and that it would "make conventional databases obsolete." We thought to ourselves, "Nonsense." Such statements demand some perspective on key dynamics that are often overlooked, which includes being mindful of where Big Data technologies are on the maturity curve, picking the right partner for the journey, understanding how it complements traditional analytic platforms (the left hand–right hand analogy from the last chapter), and considering the people component when it comes to choosing a Big Data partner. All that said, we do think Big Data is a game changer for the overall effectiveness of your data centers, because of its potential as a powerful tool in your information management repertoire.

Big Data is relatively new to many of us, but the value you want to derive out of a Big Data platform is not. Customers that intend to implement Hadoop into their enterprise environments are excited about the opportunity the MapReduce programming model offers them. While they love the idea of performing analytics on massive volumes of data that were cost prohibitive in the past, they're still enterprises with enterprise expectations and demands. For this reason, we think IBM is anything but new to this game—not to mention that we worked with Google on MapReduce projects starting back in October, 2007.

As you can imagine, IBM has tremendous assets and experience when it comes to integration solutions and ensuring that they are compliant, highly available, secure, recoverable, and provide a framework where as data flows across their information supply chain, it can be trusted (because, no one buys an IT solution because they love to run software).

Think of an artist painting a picture: A blank canvas (an IT solution) is an opportunity, and the picture you paint is the end goal—you need the right brushes and colors (and sometimes you'll mix colors to make it perfect) to paint your IT picture. With companies that only resell open source Hadoop solutions tied to services or new-to-market file systems, the discussion starts and ends at the hammer and nail needed to hang the picture on the wall. You end up having to go out and procure painting supplies and rely on your own artistic skills to paint the picture. The IBM Big Data platform is like a "color by numbers" painting kit that includes everything you need to quickly frame, paint, and hang a set of vibrant, detailed pictures with any customizations you see fit. In this kit, IBM provides everything you need, including toolsets designed for development, customization, management, and data visualization, prebuilt advanced analytic toolkits for statistics and text, and an enterprise hardening of the Hadoop runtime, all within an automated install.

IBM's world class, award winning Research arm continues to embrace and extend the Hadoop space with highly abstracted query languages, optimizations, text and machine learning analytics, and more. Other companies, especially smaller companies that leverage open source, may have some breadth in the project (as would IBM), but they typically don't have the depth to understand the collection of features that are critical to the enterprise. For example, open source has text analytics and machine learning pieces, but these aren't rounded out or as easy to use and extensible as those found in BigInsights, and this really matters to the enterprise. No doubt, for some customers, the Open Source community is all they need and IBM absolutely respects that (it's why you can solely buy a Hadoop support contract from IBM). For others who want the traditional support and delivery models, along with access to billions of dollars of investment in text and machine learning analytics, as well as other enterprise features, IBM offers its Big Data platform. IBM delivers other benefits to consider as well: 24×7 direct to engineer support, nationalized code and service in your native tongue, and more. We've got literally thousands of personnel that you can partner with to help you paint your picture. In addition, there are solutions from IBM, such as Cognos Consumer Insights, that run on BigInsights and can accelerate your Big Data projects.

When you consider all of the benefits IBM adds to a Hadoop system, you can understand why we refer to BigInsights as a *platform*. In this chapter, we cover the nontechnical details of the value IBM brings to a Big Data solution (we'll dive into the technical details in Chapter 5).

Big Data Has No Big Brother: It's Ready, but Still Young

Ask any experienced CIO, and they will be the first to tell you that in many ways the technology is the easy part. The perspective we want to offer reflects the intersection of technology and people. A good example of this is one very pragmatic question that we ask ourselves all the time: "What's worked in the warehousing space that will also be needed here?"

Note that we are not asking what technology worked; it's a broader thought than that. Although being able to create and secure data marts, enforce workload priorities, and extend the ratio of business users to developers are all grounded in technology, these are best practices that have emerged from thousands of person-years of operational experience. Here are a couple of good examples: Large IT investments are often followed by projects that get stuck in "science project mode," not because the technology failed, but because it wasn't pointed at the right problem to solve, or it could not integrate into the rest of the data center supply chain and its often complex information flows. We've also seen many initial small deployments succeed, but they are challenged to make it past the ad hoc phase because the "enterprise" part of their jobs comes calling (more on this in a bit). This can often account for the variance between the buzz around Hadoop and the lack of broad-scale notable usage. Now that sounds like a bit of an oxymoron, because Hadoop is well known and used among giants such as Twitter, Facebook, and Yahoo; but recall that all of these companies have development teams that are massive, the likes of which most Fortune 500 companies can't afford, because they are not technology companies nor do they want to be. They want to find innovative ways to accelerate their core competency businesses.

Aside from customers that have such large IT budgets they can fund a roll-your-own (RYO) environment for anything they want to do, there are a number of companies in production with Hadoop, but not in a conventional enterprise sense. For example, is data quality a requirement? Do service level agreements (SLAs) bind IT into contracts with the business sponsor? Is data secured? Are privacy policies enforced? Is the solution mission critical and therefore has survival (disaster recovery and high availability) plans with defined mean time to repair (MTTR) and recovery point objectives (RPOs) in place?

We bring up these questions because we've heard from clients that started with a "Use Hadoop but it doesn't come with the enterprise expectation bar set for other solutions in our enterprise" approach. We want to be clear about something here: We are huge fans and supporters of Hadoop and its community; however, some customers have certain needs they have asked us to address (and we think most users will end up with the same requirements). The IBM Big Data platform is about "embrace and extend." IBM embraces this open source technology (we've already detailed the long list of contributions to open source including IBM's Hadoop committers, the fact that we don't fork the code, and our commitment to maintain backwards compatibility), and extend the framework around needs voiced to us by our clients—namely analytic enrichment and some enterprise optimization features. We believe that the open source Hadoop engine, partnered with a rich ecosystem that hardens and extends it, can be a first class citizen in a business process that meets the expectations of the enterprise. After all, Hadoop isn't about speed-of-thought response times, and it's not for online transaction processing (OLTP) either; it's for batch jobs, and as we all know, batch windows are shrinking. Although businesses will extend them to gain insight never before possible, how long do you think it will be until your Hadoop project's availability and performance requirements get an "I LOVE my SLA" tattoo? It's inevitable.

The more value a Hadoop solution delivers to the enterprise, the closer it will come to the cross-hairs of criticality, and that means new expectations and new levels of production SLAs. Imagine trying to explain to your VP of Risk Management that you are unsure if your open risk positions and analytic calculations are accurate and complete. Crazy, right? Now to be fair, these challenges exist with any system and we're not saying that Hadoop isn't fantastic. However, the more popular and important it becomes within your business, the more scrutiny will be applied to the solution that runs on it. For example, you'll have audit-checking for too many open ports, you'll be asked to separate duties, you can apply the principle of least privilege to operations, and so on.

This kind of situation is more common than you would expect, and it occurs because many didn't take a step back to look at the broader context and business problem that needs to be solved. It also comes from the realities of working with young technology, and addressing this issue requires substantial innovation going forward.

IBM offers a partnership that not only gives you a platform for Big Data that flattens the time to analysis curve and addresses many enterprise needs, but it truly offers experience with critical resources and understands the importance of supporting and maintaining them. For example, the IBM Data Server drivers support billions of dollars of transactions per hour each and every day—now that's business criticality! Mix that with an innovative technology such as Hadoop, and IBM's full support for open source, and you've got a terrific opportunity.

What Can Your Big Data Partner Do for You?

So far in this chapter, we've briefly hinted at some of the capabilities that IBM offers your Big Data solution—namely, delivering a platform as opposed to a product. But behind any company you need to look at the resources it can bring to the table, how it can support you in your quest and goals, where it can support you, and whether it is working and investing in the future of the platform and enhancements that deliver more value faster. Or are they just along for the ride, giving some support for a product and not thinking based on a platform perspective, leaving you to assemble it and figure most of the enterprise challenges out for yourself.

In this section, we'll talk about some of the things IBM is doing and resources it offers that makes it a sure bet for a Big Data platform. When you look at BigInsights, you're looking at a five-year effort of more than 200 IBM research scientists with patents and award winning work. For example, IBM's General Parallel File System – Shared Nothing Cluster (GPFS-SNC) won the SC10 Storage Challenge award that is given to the most innovative storage solution in the competition.

The IBM $100 Million Big Data Investment

As a demonstration of IBM's commitment to continued innovation around the Hadoop platform, in May, 2011 it announced a $100 million investment in massive scale analytics. The key word *analytics* is something worth making note of here. Suppose multiple vendors offer some kind of Hadoop product. How many of them are rounding it out to be a platform that includes accelerators and capability around analytics? Or is that something you're left to either build from scratch yourself, or purchase and integrate separately and leverage different tools, service support contracts, code quality, and so on for your IT solutions? When you think about analytics, consider IBM SPSS and IBM Cognos assets (don't forget Unica, CoreMetrics, and so many more), alongside analytic intellectual property within Netezza or the IBM Smart Analytics System. The fact that IBM has a Business Analytics and Optimization (BAO) division speaks for itself and represents the kinds of long-term capabilities IBM will deliver for analytics in its Big Data platform. And, don't forget, to the best of our knowledge, we know of *no other* vendor that can talk *and* deliver analytics for Big Data in motion (InfoSphere Streams, or simply Streams) and Big Data at rest (BigInsights) together.

IBM can make this scale of commitment in good part because it has a century-old track record of being successful with innovation. IBM has the single largest commercial research organization on Earth, and if that's not enough, we'll finish this section with this sobering fact for you to digest about the impact a partner like IBM can have on your Big Data business goals: in the past five years, IBM has invested more than $14 billion in 24 analytics acquisitions. Today, more than 8000 IBM business consultants are dedicated to analytics and more than 200 mathematicians are developing breakthrough algorithms inside IBM Research. Now that's just for analytics; we didn't talk about the hardening of Hadoop for enterprise suitability, our committers to Apache projects (including Hadoop), and so much more. So you tell us, does this sound like the kind of player you'd like to draft for your team?

A History of Big Data Innovation

Before you read this section, we want to be clear that it's marketing information: It sounds like marketing, looks like marketing, and reads like marketing. But the thing about IBM marketing is that it's factual (we'd love to make an explicit joke here about some of our competitors, but we're sure we just did). With that said, the innovation discussed in the following sections shows that IBM has been working on and solving problems for generations, and that its research labs are typically ahead of the market and have often provided solutions for problems before they occur. As we round out the business aspect of this book, let's take a moment to reflect on the kind of partner IBM has been, is, and can be, with just a smidgen of its past innovation that can be linked to IBM's readiness to be your Big Data partner today.

The fact that IBM has a history of firsts is probably new to you: from the first traffic light timing system, to Fortran, DRAM, ATMs, UPC bar codes, RISC architecture, the PC, SQL, and XQuery, to relational database technology, and literally hundreds of other innovation assets in-between (check the source of this history at www.ibm.com/ibm100/ for a rundown of innovation that spans a century). Let's take a look at some IBM innovations over the years to see how they uniquely position IBM to be the Big Data industry leader.

1956: First Magnetic Hard Disk

IBM introduced the world's first magnetic hard disk for data storage, Random Access Method of Accounting and Control (RAMAC), offering unprecedented performance by permitting random access to any of the million characters distributed over both sides of 50 × 2-foot-diameter disks. Produced in San Jose, California, IBM's first hard disk stored about 2000 bits of data per square inch and had a purchase price of about $10,000 per megabyte. By 1997, the cost of storing a megabyte had dropped to around 10 cents. IBM is still a leader in the storage game today with innovative deduplication optimizations, automated data placement in relation to the data's utilization rates (not a bad approach when you plan to store petabytes of data), solid state disk, and more. Luckily for Big Data, the price of drives continues to drop while the capacity continues to increase; however, without the economical disk drive technology invented by IBM, Big Data would not be possible.

1970: Relational Databases

IBM scientist Ted Codd published a paper introducing the concept of relational databases. It called for information stored within a computer to be arranged in easy-to-interpret tables so that nontechnical users could access and manage large amounts of data. Today, nearly all enterprise-wide database structures are based on the IBM concept of relational databases: DB2, Informix, Netezza, Oracle, Sybase, SQL Server, and more. Your Big Data solution won't live alone; it has to integrate and will likely enhance your relational database, an area in which few other companies can claim the same kind of experience—and IBM invented it.

1971: Speech Recognition

IBM built the first operational speech recognition application that enabled engineers servicing equipment to talk to and receive spoken answers from a computer that could recognize about 5000 words. Today, IBM's ViaVoice voice recognition technology has a vocabulary of 64,000 words and a 260,000-word backup dictionary. In 1997, ViaVoice products were introduced in China and Japan. Highly customized VoiceType products are also specifically available for people working in the fields of emergency medicine, journalism, law, and radiology. Now consider speech recognition technology as it relates to the call center use case outlined in Chapter 2, and realize that IBM has intellectual property in this domain that dates back to before some readers of this book were born.

1980: RISC Architecture

IBM successfully built the first prototype computer employing RISC (Reduced Instruction Set Computer) architecture. Based on an invention by IBM scientist John Cocke in the early 1970s, the RISC concept simplified the instructions given to run computers, making them faster and more powerful. Today, RISC architecture is the basis of most enterprise servers and is widely viewed as the dominant computing architecture of the future. When you think about the computational capability required today for analytics and modeling, and what will be needed tomorrow, you're going to want a Big Data partner that owns the fabrication design of the chip that literally invented High Performance Computing (HPC) and can be found in modern-day Big Data marvels like Watson, the *Jeopardy!* champion of champions.

1988: NSFNET

IBM, working with the National Science Foundation (NSF) and our partners at MCI and Merit, designed, developed, and deployed a new high-speed network (called NSFNET) to connect approximately 200 U.S. universities and six U.S.-based supercomputer centers. The NSFNET quickly became the principal backbone of the Internet and the spark that ignited the worldwide Internet revolution. The NSFNET greatly increased the speed and capacity of the Internet (increasing the bandwidth on backbone links from 56kb/sec, to 1.5Mb/sec, to 45Mb/sec) and greatly increased the reliability and reach of the Internet to more than 50 million users in 93 countries when control of the Internet was transferred to the telecom carriers and commercial Internet Service Providers in April 1995. This expertise at Internet Scale data movement has led to significant investments in both the hardware and software required to deliver solutions capable of working at Internet Scale. In addition, a number of our cyber security and network monitoring Big Data patterns utilize packet analytics that leverage our pioneering work on the NFSNET.

1993: Scalable Parallel Systems

IBM helped pioneer the technology of joining multiple computer processors and breaking down complex, data-intensive jobs to speed their completion. This technology is used in weather prediction, oil exploration, and manufacturing. The DB2 Database Partitioning Facility (DB2 DPF)—the massively parallel processing (MPP) engine used to divide and conquer queries on a shared architecture can be found within the IBM Smart Analytics System—it has been used for decades to solve large data set problems. Although we've not yet talked about the technology in Hadoop, in Part II you're going to learn about something called MapReduce, and how its approach to parallelism (large-scale independent distributed computers working on the same problem) leverages an approach that is conceptually very similar to the DB2 DPF technology.

1996: Deep Thunder

In 1996, IBM began exploring the "business of weather," hyper-local, short-term forecasting, and customized weather modeling for clients. Now, new analytics software, and the need for organizations like cities and energy utilities to operate smarter, are changing the market climate for these services.

As Lloyd Treinish, chief scientist of the Deep Thunder program in IBM Research, explains, this approach isn't about the kind of weather reports people see on TV, but focuses on the operational problems that weather can present to businesses in very specific locales—challenges that traditional meteorology doesn't address.

For example, public weather data isn't intended to predict, with reasonable confidence, if three hours from now the wind velocity on a 10-meter diving platform will be acceptable for a high stakes competition. That kind of targeted forecasting was the challenge that IBM and the U.S. National Oceanic and Atmospheric Administration (NOAA), parent of the U.S. National Weather Service, took on in 1995.

This massive computation problem set is directly relatable to the customer work we do every day, including Vestas, which we mentioned in Chapter 2. It is also a good example of the IBM focus on analytic outcomes (derived via a platform) rather than a Big Data commitment stopping at basic infrastructure. While the computing environment here is certainly interesting, it is how the compute infrastructure was put to work that is really the innovation—exactly the same dynamic that we see in the Big Data space today.

1997: Deep Blue

The 32-node IBM RS/6000 SP supercomputer, Deep Blue, defeated World Chess Champion Garry Kasparov in the first known instance of a computer vanquishing a world champion chess player in tournament-style competition (compare this to Watson almost two decades later and there's a new inflection point with Watson being a "learning" machine). Like the preceding examples, the use of massively parallel processing is what allowed Deep Blue to be successful. Breaking up tasks into smaller subtasks and executing them in parallel across many machines is the foundation of a Hadoop cluster.

2000: Linux

In 2000, Linux received an important boost when IBM announced it would embrace Linux as strategic to its systems strategy. A year later, IBM invested $1 billion to back the Linux movement, embracing it as an operating system for IBM servers and software, stepping up to indemnify users during a period of uncertainty around its license. IBM's actions grabbed the attention of CEOs and CIOs around the globe and helped Linux become accepted by the business world. Linux is the de facto operating system for Hadoop, and you can see that your Big Data partner has more than a decade of experience in Hadoop's underlying operating system.

2004: Blue Gene

The Blue Gene supercomputer architecture was developed by IBM with a target of PFLOPS-range performance (over one quadrillion floating-point operations per second). In September 2004, an IBM Blue Gene computer broke the world record for PFLOPS. For the next four years, a computer with IBM Blue Gene architecture maintained the title of World's Fastest SuperComputer. Blue Gene has been used for a wide range of applications, including mapping the human genome, investigating medical therapies, and predicting climate trends. In 2009, American President Barack Obama awarded IBM its seventh National Medal of Technology and Innovation for the achievements of Blue Gene.

2009: The First Nationwide Smart Energy and Water Grid

The island nation of Malta turned to IBM to help mitigate its two most pressing issues: water shortages and skyrocketing energy costs. The result is a combination smart water and smart grid system that uses instrumented digital meters to monitor waste, incentive efficient resource use, deter theft, and reduce dependence on oil and processed seawater. Together, Malta and IBM are building the world's first national smart utility system. IBM has solved many of the problems you are facing today and can bring extensive domain knowledge to help you.

2009: Streams Computing

IBM announced the availability of its Streams computing software, a breakthrough data-in-motion analytics platform. Streams computing gathers multiple streams of data on the fly, using advanced algorithms to deliver nearly instantaneous analysis. Flipping the traditional data analytics strategy in which data is collected in a database to be searched or queried for answers, Streams computing can be used for complex, dynamic situations that require immediate decisions, such as predicting the spread of an epidemic or monitoring changes in the condition of premature babies. The Streams computing work was moved to IBM Software Group and is commercially available as part of the IBM Big Data platform as InfoSphere Streams (we cover it in Chapter 6). In this book, we talk about data-in-motion and data-at-rest analytics, and how you can create a cyclical system that learns and delivers unprecedented vision; this is something we believe only IBM can deliver as part of a partnership at this time. You might be wondering just what kind of throughput Streams can sustain while running analytics. In one customer environment, Streams analyzed 500,000 call detail records (CDR) per second (a kind of detail record for cellular communications), processing over 6 billion CDRs per day and over 4 PBs of data per year!

2009: Cloud

IBM's comprehensive capabilities make the Enterprise Cloud promise a reality. IBM has helped thousands of clients reap the benefits of cloud computing: With over 2000 private cloud engagements in 2010 alone, IBM manages billions of cloud-based transactions every day with millions of cloud users. IBM itself is using cloud computing extensively and experiencing tremendous benefits, such as accelerated deployment of innovative ideas and more than $15 million a year in savings from their development. Yet obtaining substantial benefits to address today's marketplace realities is not a matter of simply implementing cloud capabilities—but of how organizations strategically utilize new ways to access and mix data. Too often this vast potential is unmet because cloud technology is being used primarily to make IT easier, cheaper, and faster. IBM believes that the Cloud needs to be about transformation. While it obviously includes how IT is delivered, the vision is extended to think about what insight is delivered; doing this requires both the platform capabilities to handle the volume, variety, and velocity of the data, and more importantly, being able to build and deploy the analytics required that result in transformational capabilities.

2010: GPFS SNC

Initially released in 1998, the IBM General Parallel File System (GPFS) is a high-performance POSIX-compliant shared-disk clustered file system that runs on a storage area network (SAN). Today, GPFS is used by many supercomputers, DB2 pureScale, many Oracle RAC implementations, and more. GPFS provides concurrent high-speed file access to applications executing on multiple nodes of clusters through its ability to stripe blocks of data across multiple disks and by being able to read them in parallel. In addition, GPFS provides high availability, disaster recovery, security, hierarchal storage management, and more. GPFS was extended to run on shared-nothing clusters (known as GFPS-SNC) and took the SC10 Storage Challenge 2010 award for the most innovative storage solution: "It is designed to reliably store petabytes to exabytes of data while processing highly parallel applications like Hadoop analytics twice as fast as competing solutions." A well-known enterprise class file system extended for Hadoop suitability is compelling for many organizations.

2011: Watson

IBM's Watson leverages leading-edge Question-Answering (QA) technology, allowing a computer to process and understand natural language. Watson also implemented a deep-rooted learning behavior that understood previous correct and incorrect decisions, and it could even apply risk analysis to future decisions and domain knowledge. Watson incorporates massively parallel analytical capabilities to emulate the human mind's ability to understand the actual meaning behind words, distinguish between relevant and irrelevant content, and, ultimately, demonstrate confidence to deliver precise final answers. In February 2011, Watson made history by not only being the first computer to compete against humans on television's venerable quiz show, *Jeopardy!*, but by achieving a landslide win over champions Ken Jennings and Brad Rutter. Decision Augmentation on diverse knowledge sets is an important application of Big Data technology, and Watson's use of Hadoop to store and pre-process its corpus of knowledge is a foundational capability for BigInsights going forward. Here again, if you focus your vendor selection merely on supporting Hadoop, you miss the key value—the discovery of understanding and insight—rather than just processing data.

IBM Research: A Core Part of InfoSphere BigInsights Strategy

Utilizing IBM Research's record of innovation has been a deliberate part of the IBM Big Data strategy and platform. In addition to Streams, IBM started BigInsights in the IBM Research labs and moved it to IBM Software Group (SWG) more than a year prior to its general availability.

Some of the IBM Research inventions such as the *Advanced Text Analytics Toolkit* (previously known as SystemT) and *Intelligent Scheduler* (provides workload governance above and beyond what Hadoop offers—it was previously known as *FLEX*) were shipped with the first BigInsights release. Other innovations such as GPFS-SNC (synonymous for more than 12 years with enterprise performance and availability), *Adaptive MapReduce*, and the *Machine Learning Toolkit* (you may have heard it previously referred to as System ML) are either available today, or are soon to be released. (You'll notice the BigInsights development teams have adopted a start-up mentality for feature delivery—they come quickly and often as opposed to traditional software releases.) We cover all of these technologies in Part II.

IBM Research is the fundamental engine that is driving Big Data analytics and the hardening of the Hadoop ecosystem. And, of course, the BigInsights effort is not only driven by IBM Research: Our Big Data development teams in Silicon Valley, India, and China have taken the technologies from IBM Research and further enhanced them, which resulted in our first commercial releases based on substantial input from both external and internal customers.

IBM's Internal Code-mart and Big Data

Hadoop is an Apache top-level project. One thing that not many people outside of IBM know is that IBM has its own internal version of the Apache model where teams can leverage other teams' software code and projects and use them within their own solutions, and then contribute enriched code back to the central IBM community. For example, DB2 pureScale leverages technologies found in Tivoli System Automation, GPFS, and HACMP. When DB2 pureScale was developed, a number of enhancements were put into these technologies, which materialized as enhancements in their own respective commercially available products. Although this sharing has been going on for a long time, the extent and speed as it relates to emerging technologies has been dramatically accelerated and will be a key part of IBM's Big Data partnership and journey.

Just as IBM did with its innovative Mashup technology that was created by an Information Management and Lotus partnership, but quickly leveraged by IBM Enterprise Content Management, Cognos, WebSphere, and Tivoli, the IBM Big Data teams are already seeing similar code sharing around their Hadoop efforts. IBM Cognos Consumer Insight (CCI) is a good example of a generally available product that got to market more quickly because of this sharing within the IBM Big Data portfolio. CCI runs on BigInsights (as do other IBM products) and enables marketing professionals to be more precise, agile, and responsive to customer demands and opinions expressed through social media by analyzing large volumes of publicly available Internet content. CCI utilizes BigInsights to collect, store, and perform the foundational analytic processes needed on this content, and augments that with application-level social media analytics and visualization. BigInsights can utilize the CCI collected data for follow-on analytic jobs, including using internal enterprise data sources to find the correlations between the CCI identified behavior and what the enterprise did to influence or drive that behavior (coupons, merchandising mix, new products, and so on).

We've encouraged this level of code sharing for several reasons, but perhaps the most important of those is that the diversity of usage brings along a diversity of needs and domain expertise. We recognize we are on a journey here, so enlisting as many guides as possible helps. Of course, when on a journey it helps to pick experienced guides, since learning to skydive from a scuba instructor may not end well. Relevant expertise matters.

Domain Expertise Matters

There are hundreds of examples of deep domain expertise being applied to solving previously unsolvable problems with Big Data. This section shows two examples, one from the media industry and one from the energy industry.

We recently used BigInsights to help a media company qualify how often its media streams were being distributed without permission. The answer was a lot—much more, in fact, than it had expected. It was a successful use of the technology, but did it solve the business problem? No—and in fact this represented only the start of the business problem. The easy response would have been to try and clamp down on those "without expressed written consent," but would that have been the right business decision? Not necessarily, since this audience turns out to be under served with thirst for consumption, and although using copyright materials without the owner's permission is clearly bad, this was an opportunity in disguise: the market was telling the firm that a whole new geography was interested in their copyrighted assets—which represented a new opportunity they previously didn't see. Reaching the right decision from a business strategy perspective is rarely, if ever, a technology decision. This is where domain expertise is critical to ensure the technology is applied appropriately in the broader business context. A good example of this applied expertise is IBM's Smart Planet work, which implicitly is the application of domain expertise to ever larger and diverse data sets.

Let's take a look at the Energy sector. It is estimated that by 2035, the global demand for energy will rise by nearly 50 percent, and while renewable energy sources will start to make a significant contribution towards this increased demand, conventional oil and gas will still need to make up a full 50 percent of that total increase in demand. This will be no small accomplishment given that the oil and gas reserves needing to be accessed are increasingly remote. Finding, accessing, refining, and transporting those reserves—profitably and safely—is demanding new ways of understanding how all the pieces come together. The amount and diversity of the data generated in the oil and gas production cycle is staggering. Each well can have more than 20,000 individual sensors generating multiple terabytes per day (or much more) per well. Simply storing the output from a field of related wells can be a 10 petabytes (or more) yearly challenge, now add to that the compute requirements to correlate behavior across those wells. While there is clearly potential to harvest important understanding from all of this data, knowing where to start and how to apply it is critical.

Of course, once available for consumption, optimizing how energy is distributed is a natural follow-on. IBM Smart Grid is a good example of the intersection of domain expertise meeting Big Data. IBM is helping utilities add a layer of digital intelligence to their grids. These smart grids use sensors, meters, digital controls, and analytic tools to automate, monitor, and control the two-way flow of energy across operations—from power plant to plug. A power company can optimize grid performance, prevent outages, restore outages faster, and allow consumers to manage energy usage right down to an individual networked appliance.

The IBM Big Data platform allows for the collection of all the data needed to do this, but more importantly, the platform's analytic engines can find the correlation of conditions that provide new awareness into how the grid can be optimally run. As you can imagine, the ability to store events is important, but being able to understand and link events in this domain is critical. We like to refer to this as finding signals in the noise. The techniques, analytic engines, and domain expertise IBM has developed in this space are equally applicable in understanding 360-degree views of a customer, especially when social media is part of the mix.

Part II

Big Data:
From the Technology
Perspective

4

All About Hadoop: The Big Data Lingo Chapter

It should be evident now that you've read Part I, but we have a hunch you already figured this out before picking up this book—there are mountains of untapped potential in our information. Until now, it's been too cost prohibitive to analyze these massive volumes. Of course, there's also been a staggering opportunity cost associated with not tapping into this information, as the potential of this yet-to-be-analyzed information is near-limitless. And we're not just talking the ubiquitous "competitive differentiation" marketing slogan here; we're talking innovation, discovery, association, and pretty much anything else that could make the way you work tomorrow very different, with even more tangible results and insight, from the way you work today.

People and organizations have attempted to tackle this problem from many different angles. Of course, the angle that is currently leading the pack in terms of popularity for massive data analysis is an open source project called *Hadoop* that is shipped as part of the IBM InfoSphere BigInsights (BigInsights) platform. Quite simply, BigInsights embraces, hardens, and extends the Hadoop open source framework with enterprise-grade security, governance, availability, integration into existing data stores, tooling that simplifies and improves developer productivity, scalability, analytic toolkits, and more.

When we wrote this book, we thought it would be beneficial to include a chapter about Hadoop itself, since BigInsights is (and will always be) based on the nonforked core Hadoop distribution, and backwards compatibility

53

with the Apache Hadoop project will always be maintained. In short, applications written for Hadoop will always run on BigInsights. This chapter isn't going to make you a Hadoop expert by any means, but after reading it, you'll understand the basic concepts behind the core Hadoop technology, and you might even sound really smart with the nontechies at the water cooler. If you're new to Hadoop, this chapter's for you.

Just the Facts: The History of Hadoop

Hadoop (http://hadoop.apache.org/) is a top-level Apache project in the Apache Software Foundation that's written in Java. For all intents and purposes, you can think of Hadoop as a computing environment built on top of a distributed clustered file system that was designed specifically for very large-scale data operations.

Hadoop was inspired by Google's work on its Google (distributed) File System (GFS) and the MapReduce programming paradigm, in which work is broken down into mapper and reducer tasks to manipulate data that is stored across a cluster of servers for massive parallelism. MapReduce is not a new concept (IBM teamed up with Google in October 2007 to do some joint university research on MapReduce and GFS for large-scale Internet problems); however, Hadoop has made it practical to be applied to a much wider set of use cases. Unlike transactional systems, Hadoop is designed to scan through large data sets to produce its results through a highly scalable, distributed batch processing system. Hadoop is not about speed-of-thought response times, real-time warehousing, or blazing transactional speeds; *it is* about discovery and making the once near-impossible possible from a scalability and analysis perspective. The Hadoop methodology is built around a function-to-data model as opposed to data-to-function; in this model, because there is so much data, the analysis programs are sent to the data (we'll detail this later in this chapter).

Hadoop is quite the odd name (and you'll find a lot of odd names in the Hadoop world). Read any book on Hadoop today and it pretty much starts with the name that serves as this project's mascot, so let's start there too. Hadoop is actually the name that creator Doug Cutting's son gave to his stuffed toy elephant. In thinking up a name for his project, Cutting was apparently looking for something that was easy to say and stands for nothing in particular, so the name of his son's toy seemed to make perfect sense.

Cutting's naming approach has kicked off a wild collection of names (as you will soon find out), but to be honest, we like it. (We reflected among ourselves about some of the names associated with our kids' toys while we wrote this book, and we're glad Cutting dubbed this technology and not us; Pinky and Squiggles don't sound like good choices.)

Hadoop is generally seen as having two parts: a file system (the *Hadoop Distributed File System*) and a programming paradigm (*MapReduce*)—more on these in a bit. One of the key components of Hadoop is the redundancy built into the environment. Not only is the data redundantly stored in multiple places across the cluster, but the programming model is such that failures *are expected* and are resolved automatically by running portions of the program on various servers in the cluster. Due to this redundancy, it's possible to distribute the data and its associated programming across a very large cluster of commodity components. It is well known that commodity hardware components will fail (especially when you have very large numbers of them), but this redundancy provides fault tolerance and a capability for the Hadoop cluster to heal itself. This allows Hadoop to scale out workloads across large clusters of inexpensive machines to work on Big Data problems.

There are a number of Hadoop-related projects, and some of these we cover in this book (and some we don't, due to its size). Some of the more notable Hadoop-related projects include: Apache Avro (for data serialization), Cassandra and HBase (databases), Chukwa (a monitoring system specifically designed with large distributed systems in mind), Hive (provides ad hoc SQL-like queries for data aggregation and summarization), Mahout (a machine learning library), Pig (a high-level Hadoop programming language that provides a data-flow language and execution framework for parallel computation), ZooKeeper (provides coordination services for distributed applications), and more.

Components of Hadoop

The Hadoop project is comprised of three pieces: *Hadoop Distributed File System (HDFS)*, the *Hadoop MapReduce* model, and *Hadoop Common*. To understand Hadoop, you must understand the underlying infrastructure of the file system and the MapReduce programming model. Let's first talk about Hadoop's file system, which allows applications to be run across multiple servers.

The Hadoop Distributed File System

To understand how it's possible to scale a Hadoop cluster to hundreds (and even thousands) of nodes, you have to start with HDFS. Data in a Hadoop cluster is broken down into smaller pieces (called *blocks*) and distributed throughout the cluster. In this way, the map and reduce functions can be executed on smaller subsets of your larger data sets, and this provides the scalability that is needed for Big Data processing.

The goal of Hadoop is to use commonly available servers in a very large cluster, where each server has a set of inexpensive internal disk drives. For higher performance, MapReduce tries to assign workloads to these servers where the data to be processed is stored. This is known as *data locality*. (It's because of this principle that using a storage area network (SAN), or network attached storage (NAS), in a Hadoop environment is not recommended. For Hadoop deployments using a SAN or NAS, the extra network communication overhead can cause performance bottlenecks, especially for larger clusters.) Now take a moment and think of a 1000-machine cluster, where each machine has three internal disk drives; then consider the failure rate of a cluster composed of 3000 *inexpensive* drives + 1000 *inexpensive* servers!

We're likely already on the same page here: The component mean time to failure (MTTF) you're going to experience in a Hadoop cluster is likely analogous to a zipper on your kid's jacket: it's going to fail (and poetically enough, zippers seem to fail only when you really need them). The cool thing about Hadoop is that the reality of the MTTF rates associated with inexpensive hardware is actually well understood (a design point if you will), and part of the strength of Hadoop is that it has built-in fault tolerance and fault compensation capabilities. This is the same for HDFS, in that data is divided into blocks, and copies of these blocks are stored on other servers in the Hadoop cluster. That is, an individual file is actually stored as smaller blocks that are replicated across multiple servers in the entire cluster.

Think of a file that contains the phone numbers for everyone in the United States; the people with a last name starting with *A* might be stored on server 1, *B* on server 2, and so on. In a Hadoop world, pieces of this phonebook would be stored across the cluster, and to reconstruct the entire phonebook, your program would need the blocks from every server in the cluster. To achieve availability as components fail, HDFS replicates these smaller pieces (see Figure 4-1) onto two additional servers by default. (This redundancy can

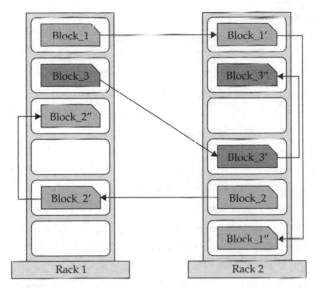

Figure 4-1   *An example of how data blocks are written to HDFS. Notice how (by default) each block is written three times and at least one block is written to a different server rack for redundancy.*

be increased or decreased on a per-file basis or for a whole environment; for example, a development Hadoop cluster typically doesn't need any data redundancy.) This redundancy offers multiple benefits, the most obvious being higher availability. In addition, this redundancy allows the Hadoop cluster to break work up into smaller chunks and run those jobs on all the servers in the cluster for better scalability. Finally, you get the benefit of data locality, which is critical when working with large data sets. We detail these important benefits later in this chapter.

A data file in HDFS is divided into blocks, and the default size of these blocks for Apache Hadoop is 64 MB. For larger files, a higher block size is a good idea, as this will greatly reduce the amount of metadata required by the NameNode. The expected workload is another consideration, as nonsequential access patterns (random reads) will perform more optimally with a smaller block size. In BigInsights, the default block size is 128 MB, because in the experience of IBM Hadoop practitioners, the most common deployments involve larger files and workloads with sequential reads. This is a much larger block size than is used with other environments—for example, typical file systems have an on-disk block size of 512 bytes, whereas relational databases typically store data blocks in sizes ranging from 4 KB to 32 KB. Remember that Hadoop was designed to scan through very large data sets, so it makes sense for it to use a very large block size so that each server can work on a larger chunk of data at the same time. Coordination across a cluster has significant overhead, so the ability to process large chunks of work locally without sending data to other nodes helps improve both performance and the *overhead to real work* ratio. Recall that each data block is stored by default on three different servers; in Hadoop, this is implemented by HDFS working behind the scenes to make sure at least two blocks are stored on a separate server rack to improve reliability in the event you lose an entire rack of servers.

All of Hadoop's data placement logic is managed by a special server called *NameNode*. This NameNode server keeps track of all the data files in HDFS, such as where the blocks are stored, and more. All of the NameNode's information is stored in memory, which allows it to provide quick response times to storage manipulation or read requests. Now, we know what you're thinking: If there is only one NameNode for your entire Hadoop cluster, you need to be aware that storing this information in memory creates a single point of failure (SPOF). For this reason, we *strongly* recommend that the server components you choose for the NameNode be much more robust than the rest of the servers in your Hadoop cluster to minimize the possibility of failures. In addition, we also strongly recommend that you have a regular backup process for the cluster metadata stored in the NameNode. Any data loss in this metadata will result in a permanent loss of corresponding data in the cluster. When this book was written, the next version of Hadoop (version 0.21) was to include the capability to define a BackupNode, which can act as a cold standby for the NameNode.

Figure 4-1 represents a file that is made up of three data blocks, where a data block (denoted as block_n) is replicated on two additional servers (denoted by block_n' and block_n''). The second and third replicas are stored on a separate physical rack, on separate nodes for additional protection.

We're detailing how HDFS stores data blocks to give you a brief introduction to this Hadoop component. The great thing about the Hadoop MapReduce application framework is that, unlike prior grid technologies, the developer doesn't have to deal with the concepts of the NameNode and where data is stored—Hadoop does that for you. When you fire off a Hadoop job and the application has to read data and starts to work on the programmed MapReduce tasks, Hadoop will contact the NameNode, find the servers that hold the parts of the data that need to be accessed to carry out the job, and then send your application to run locally on those nodes. (We cover the details of MapReduce in the next section.) Similarly, when you create a file, HDFS will automatically communicate with the NameNode to allocate storage on specific servers and perform the data replication. It's important to note that when you're working with data, there's no need for your MapReduce code to directly reference the NameNode. Interaction with the NameNode is mostly done when the jobs are scheduled on various servers in the Hadoop cluster. This greatly reduces communications to the NameNode during job execution, which helps to improve scalability of the solution. In summary, the NameNode deals with cluster metadata describing where files are stored; actual data being processed by MapReduce jobs never flows through the NameNode.

In this book, we talk about how IBM brings enterprise capability to Hadoop, and this is one specific area where IBM uses its decades of experience and research to leverage its ubiquitous enterprise IBM General Parallel File System (GPFS) to alleviate these concerns. GPFS initially only ran on SAN technologies. In 2009, GPFS was extended to run on a shared nothing cluster (known as GPFS-SNC) and is intended for use cases like Hadoop. GFPS-SNC provides many advantages over HDFS, and one of them addresses the aforementioned NameNode issue. A Hadoop runtime implemented within GPFS-SNC does not have to contend with this particular SPOF issue. GPFS-SNC allows you to build a more reliable Hadoop cluster (among other benefits such as easier administration and performance).

In addition to the concerns expressed about a single NameNode, some clients have noted that HDFS is not a Portable Operating System Interface for UNIX (POSIX)–compliant file system. What this means is that almost all of the familiar commands you might use in interacting with files (copying files, deleting files, writing to files, moving files, and so on) are available in a different form with HDFS (there are syntactical differences and, in some cases, limitations in functionality). To work around this, you either have to write your own Java applications to perform some of the functions, or train your IT staff to learn the different HDFS commands to manage and manipulate files in the file system. We'll go into more detail on this topic later in the chapter, but here we want you to note that this is yet another "Enterprise-rounding" that BigInsights offers to Hadoop environments for Big Data processing. GPFS-SNC is fully compliant with the IEEE-defined POSIX standard that defines an API, shell, and utility interfaces that provide compatibility across different flavors of UNIX (such as AIX, Apple OSX, and HP-UX).

The Basics of MapReduce

MapReduce is the heart of Hadoop. It is this programming paradigm that allows for massive scalability across hundreds or thousands of servers in a Hadoop cluster. The MapReduce concept is fairly simple to understand for those who are familiar with clustered scale-out data processing solutions. For people new to this topic, it can be somewhat difficult to grasp, because it's not typically something people have been exposed to previously. If you're new to Hadoop's MapReduce jobs, don't worry: we're going to describe it in a way that gets you up to speed quickly.

The term *MapReduce* actually refers to two separate and distinct tasks that Hadoop programs perform. The first is the map job, which takes a set of data and converts it into another set of data, where individual elements are broken down into *tuples* (key/value pairs). The reduce job takes the output from a map as input and combines those data tuples into a smaller set of tuples. As the sequence of the name MapReduce implies, the reduce job is always performed after the map job.

Let's look at a simple example. Assume you have five files, and each file contains two columns (a *key* and a *value* in Hadoop terms) that represent a city and the corresponding temperature recorded in that city for the various measurement days. Of course we've made this example very simple so it's easy to follow. You can imagine that a real application won't be quite so simple, as it's likely to contain millions or even billions of rows, and they might not be neatly formatted rows at all; in fact, no matter how big or small the amount of data you need to analyze, the key principles we're covering here remain the same. Either way, in this example, *city is the key* and *temperature is the value*.

The following snippet shows a sample of the data from one of our test files (incidentally, in case the temperatures have you reaching for a hat and gloves, they are in Celsius):

```
Toronto, 20
Whitby, 25
New York, 22
Rome, 32
Toronto, 4
Rome, 33
New York, 18
```

Out of all the data we have collected, we want to find the maximum tem-perature for *each* city across all of the data files (note that each file might have the same city represented multiple times). Using the MapReduce framework, we can break this down into five map tasks, where each mapper works on one of the five files and the mapper task goes through the data and returns the maximum temperature for each city. For example, the results produced from one mapper task for the data above would look like this:

```
(Toronto, 20)  (Whitby, 25)  (New York, 22)  (Rome, 33)
```

Let's assume the other four mapper tasks (working on the other four files not shown here) produced the following intermediate results:

```
(Toronto, 18)  (Whitby, 27)  (New York, 32)  (Rome, 37)
(Toronto, 32)  (Whitby, 20)  (New York, 33)  (Rome, 38)
(Toronto, 22)  (Whitby, 19)  (New York, 20)  (Rome, 31)
(Toronto, 31)  (Whitby, 22)  (New York, 19)  (Rome, 30)
```

All five of these output streams would be fed into the reduce tasks, which combine the input results and output a single value for each city, producing a final result set as follows:

```
(Toronto, 32)  (Whitby, 27)  (New York, 33)  (Rome, 38)
```

As an analogy, you can think of map and reduce tasks as the way a cen-sus was conducted in Roman times, where the census bureau would dis-patch its people to each city in the empire. Each census taker in each city would be tasked to count the number of people in that city and then return their results to the capital city. There, the results from each city would be reduced to a single count (sum of all cities) to determine the overall popula-tion of the empire This *mapping* of people to cities, in parallel, and then com-bining the results (*reducing*) is much more efficient than sending a single per-son to count every person in the empire in a serial fashion.

In a Hadoop cluster, a MapReduce program is referred to as a *job*. A job is executed by subsequently breaking it down into pieces called *tasks*.

An application submits a job to a specific node in a Hadoop cluster, which is running a daemon called the *JobTracker*. The JobTracker communicates with the NameNode to find out where all of the data required for this job exists across the cluster, and then breaks the job down into map and reduce tasks for each node to work on in the cluster. These tasks are scheduled on the nodes in the cluster where the data exists. Note that a node might be given a task for which the data needed by that task is not local to that node. In such a case, the node would have to ask for the data to be sent across the network interconnect to perform its task. Of course, this isn't very efficient, so the JobTracker tries to avoid this and attempts to schedule tasks where the data is stored. This is the concept of data locality we introduced earlier, and it is critical when working with large volumes of data. In a Hadoop cluster, a set of continually running daemons, referred to as *TaskTracker* agents, monitor the status of each task. If a task fails to complete, the status of that failure is reported back to the JobTracker, which will then reschedule that task on another node in the cluster. (You can dictate how many times the task will be attempted before the entire job gets cancelled.)

Figure 4-2 shows an example of a MapReduce flow. You can see that multiple reduce tasks can serve to increase the parallelism and improve the overall performance of the job. In the case of Figure 4-2, the output of the map tasks must be directed (by key value) to the appropriate reduce task. If we apply our maximum temperature example to this figure, all of the records that have a key value of Toronto must be sent to the same reduce task to

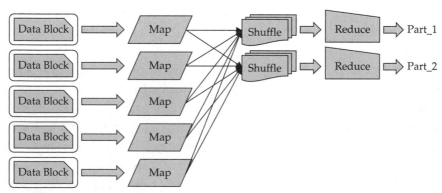

Figure 4-2   *The flow of data in a simple MapReduce job*

produce an accurate result (one reducer must be able to see all of the temperatures for Toronto to determine the maximum for that city). This directing of records to reduce tasks is known as a *Shuffle*, which takes input from the map tasks and directs the output to a specific reduce task. Hadoop gives you the option to perform local aggregation on the output of each map task before sending the results off to a reduce task through a local aggregation called a *Combiner* (but it's not shown in Figure 4-2). Clearly more work and overhead are involved in running multiple reduce tasks, but for very large datasets, having many reducers can improve overall performance.

All MapReduce programs that run natively under Hadoop are written in Java, and it is the Java Archive file (jar) that's distributed by the JobTracker to the various Hadoop cluster nodes to execute the map and reduce tasks. For further details on MapReduce, you can review the Apache Hadoop documentation's tutorial that leverages the ubiquitous Hello World programming language equivalent for Hadoop: WordCount. WordCount is a simple to understand example with all of the Java code needed to run the samples.

Of course, if you're looking for the fastest and easiest way to get up and running with Hadoop, check out BigDataUniversity.com and download Info-Sphere BigInsights Basic Edition (www.ibm.com/software/data/infosphere/biginsights/basic.html). It's got some of the great IBM add-on capabilities (for example, the whole up and running experience is completely streamlined for you, so you get a running Hadoop cluster in the same manner that you'd see in any commercial software) and more. Most importantly, it's 100 percent free, and you can optionally buy a support contract for BigInsights' Basic Edition. Of course, by the time you are finished reading this book, you'll have a complete grasp as to how IBM InfoSphere BigInsights Enterprise Edition embraces and extends the Hadoop stack to provide the same capabilities expected from other enterprise systems.

Hadoop Common Components

The Hadoop Common Components are a set of libraries that support the various Hadoop subprojects. Earlier in this chapter, we mentioned some of these components in passing. In this section, we want to spend time discussing the file system shell. As mentioned (and this is a really important point, which is why we are making note of it again), HDFS *is not* a POSIX-compliant file system, which means you can't interact with it as you would a Linux- or UNIX-based file system. To interact with files in HDFS, you need to use the `/bin/hdfs dfs <args>` file system shell command interface, where `args` represents the command arguments you want to use on files in the file system.

Here are some examples of HDFS shell commands:

`cat`	Copies the file to standard output (`stdout`).
`chmod`	Changes the permissions for reading and writing to a given file or set of files.
`chown`	Changes the owner of a given file or set of files.
`copyFromLocal`	Copies a file from the local file system into HDFS.
`copyToLocal`	Copies a file from HDFS to the local file system.
`cp`	Copies HDFS files from one directory to another.
`expunge`	Empties all of the files that are in the trash. When you delete an HDFS file, the data is not actually gone (think of your MAC or Windows-based home computers, and you'll get the point). Deleted HDFS files can be found in the trash, which is automatically cleaned at some later point in time. If you want to empty the trash immediately, you can use the `expunge` argument.
`ls`	Displays a listing of files in a given directory.
`mkdir`	Creates a directory in HDFS.
`mv`	Moves files from one directory to another.
`rm`	Deletes a file and sends it to the trash. If you want to skip the trash process and delete the file from HDFS on the spot, you can use the `-skiptrash` option of the `rm` command.

Application Development in Hadoop

As you probably inferred from the preceding section, the Hadoop platform can be a powerful tool for manipulating extremely large data sets. However, the core Hadoop MapReduce APIs are primarily called from Java, which requires skilled programmers. In addition, it is even more complex for programmers to develop and maintain MapReduce applications for business applications that require long and pipelined processing.

If you've been around programming long enough, you'll find history has a way of repeating itself. For example, we often cite XML as "The Revenge of IMS" due to its hierarchal nature and retrieval system. In the area of computer language development, just as assembler gave way to structured programming languages and then to the development of 3GL and 4GL languages, so too goes the world of Hadoop application development languages. To abstract some of the complexity of the Hadoop programming model, several application development languages have emerged that run on top of Hadoop. In this section, we cover three of the more popular ones, which admittedly sound like we're at a zoo: *Pig, Hive,* and *Jaql* (by the way, we'll cover *ZooKeeper* in this chapter, too).

Pig and PigLatin

Pig was initially developed at Yahoo! to allow people using Hadoop to focus more on analyzing large data sets and spend less time having to write mapper and reducer programs. Like actual pigs, who eat almost anything, the Pig programming language is designed to handle any kind of data—hence the name! Pig is made up of two components: the first is the language itself, which is called *PigLatin* (yes, people naming various Hadoop projects do tend to have a sense of humor associated with their naming conventions), and the second is a runtime environment where PigLatin programs are executed. Think of the relationship between a Java Virtual Machine (JVM) and a Java application. In this section, we'll just refer to the whole entity as *Pig*.

Let's first look at the programming language itself so that you can see how it's significantly easier than having to write mapper and reducer programs. The first step in a Pig program is to LOAD the data you want to manipulate from HDFS. Then you run the data through a set of transformations (which, under the covers, are translated into a set of mapper and reducer tasks). Finally, you DUMP the data to the screen or you STORE the results in a file somewhere.

LOAD

As is the case with all the Hadoop features, the objects that are being worked on by Hadoop are stored in HDFS. In order for a Pig program to access this data, the program must first tell Pig what file (or files) it will use, and that's done through the LOAD 'data_file' command (where 'data_file' specifies either an HDFS file or directory). If a directory is specified, all the files in that directory will be loaded into the program. If the data is stored in a file format that is not natively accessible to Pig, you can optionally add the USING function to the LOAD statement to specify a user-defined function that can read in and interpret the data.

TRANSFORM

The transformation logic is where all the data manipulation happens. Here you can FILTER out rows that are not of interest, JOIN two sets of data files, GROUP data to build aggregations, ORDER results, and much more. The following is an example of a Pig program that takes a file composed of Twitter feeds, selects only those tweets that are using the en (English) iso_language code, then groups them by the user who is tweeting, and displays the sum of the number of retweets of that user's tweets.

```
L  = LOAD 'hdfs//node/tweet_data';
FL = FILTER L BY iso_language_code  EQ 'en';
G  = GROUP FL BY from_user;
RT = FOREACH G GENERATE group, SUM(retweets);
```

DUMP and STORE

If you don't specify the DUMP or STORE command, the results of a Pig program are not generated. You would typically use the DUMP command, which sends the output to the screen, when you are debugging your Pig programs. When you go into production, you simply change the DUMP call to a STORE call so that any results from running your programs are stored in a file for further processing or analysis. Note that you can use the DUMP command anywhere in your program to dump intermediate result sets to the screen, which is very useful for debugging purposes.

Now that we've got a Pig program, we need to have it run in the Hadoop environment. Here is where the Pig runtime comes in. There are three ways to run a Pig program: embedded in a script, embedded in a Java program, or from the Pig command line, called *Grunt* (which is of course the sound a pig makes—we told you that the Hadoop community has a lighter side).

No matter which of the three ways you run the program, the Pig runtime environment translates the program into a set of map and reduce tasks and runs them under the covers on your behalf. This greatly simplifies the work associated with the analysis of large amounts of data and lets the developer focus on the analysis of the data rather than on the individual map and reduce tasks.

Hive

Although Pig can be quite a powerful and simple language to use, the downside is that it's something new to learn and master. Some folks at Facebook developed a runtime Hadoop support structure that allows anyone who is already fluent with SQL (which is commonplace for relational database developers) to leverage the Hadoop platform right out of the gate. Their creation, called *Hive*, allows SQL developers to write Hive Query Language (HQL) statements that are similar to standard SQL statements; now you should be aware that HQL is limited in the commands it understands, but it is still pretty useful. HQL statements are broken down by the Hive service into MapReduce jobs and executed across a Hadoop cluster.

For anyone with a SQL or relational database background, this section will look very familiar to you. As with any database management system (DBMS), you can run your Hive queries in many ways. You can run them from a command line interface (known as the *Hive shell*), from a Java Database Connectivity (JDBC) or Open Database Connectivity (ODBC) application leveraging the Hive JDBC/ODBC drivers, or from what is called a *Hive Thrift Client*. The Hive Thrift Client is much like any database client that gets installed on a user's client machine (or in a middle tier of a three-tier architecture): it communicates with the Hive services running on the server. You can use the Hive Thrift Client within applications written in C++, Java, PHP, Python, or Ruby (much like you can use these client-side languages with embedded SQL to access a database such as DB2 or Informix). The following shows an example of creating a table, populating it, and then querying that table using Hive:

```
CREATE TABLE Tweets(from_user STRING, userid BIGINT, tweettext STRING,
    retweets INT)
 COMMENT 'This is the Twitter feed table'
 STORED AS SEQUENCEFILE;
LOAD DATA INPATH 'hdfs://node/tweetdata' INTO TABLE TWEETS;
SELECT from_user, SUM(retweets)
 FROM TWEETS
 GROUP BY from_user;
```

As you can see, Hive looks very much like traditional database code with SQL access. However, because Hive is based on Hadoop and MapReduce operations, there are several key differences. The first is that Hadoop is intended for long sequential scans, and because Hive is based on Hadoop, you can expect queries to have a very high latency (many minutes). This means that Hive would not be appropriate for applications that need very fast response times, as you would expect with a database such as DB2. Finally, Hive is read-based and therefore not appropriate for transaction processing that typically involves a high percentage of write operations.

Jaql

Jaql is primarily a query language for JavaScript Object Notation (JSON), but it supports more than just JSON. It allows you to process both structured and nontraditional data and was donated by IBM to the open source community (just one of many contributions IBM has made to open source). Specifically, Jaql allows you to select, join, group, and filter data that is stored in HDFS, much like a blend of Pig and Hive. Jaql's query language was inspired by many programming and query languages, including Lisp, SQL, XQuery, and Pig. Jaql is a functional, declarative query language that is designed to process large data sets. For parallelism, Jaql rewrites high-level queries, when appropriate, into "low-level" queries consisting of MapReduce jobs.

Before we get into the Jaql language, let's first look at the popular data interchange format known as JSON, so that we can build our Jaql examples on top of it. Application developers are moving in large numbers towards JSON as their choice for a data interchange format, because it's easy for humans to read, and because of its structure, it's easy for applications to parse or generate.

JSON is built on top of two types of structures. The first is a collection of name/value pairs (which, as you learned earlier in the "The Basics of MapReduce" section, makes it ideal for data manipulation in Hadoop, which works on key/value pairs). These name/value pairs can represent anything since they are simply text strings (and subsequently fit well into existing models) that could represent a record in a database, an object, an associative array, and more. The second JSON structure is the ability to create an ordered list of values much like an array, list, or sequence you might have in your existing applications.

An object in JSON is represented as { `string : value` }, where an array can be simply represented by [`value, value, ...`], where value can be a string, number, another JSON object, or another JSON array. The following shows an example of a JSON representation of a Twitter feed (we've removed many of the fields that are found in the tweet syntax to enhance readability):

```
results: [
{
created_at: "Thurs, 14 Jul 2011 09:47:45 +0000"
from_user: "eatonchris"
geo: {
            coordinates: [
                43.866667
                78.933333
                ]
            type: "Point"
    }
iso_language_code: "en"
text: " Reliance Life Insurance migrates from #Oracle
to #DB2 and cuts costs in half. Read what they say
about their migration http://bit.ly/pP7vaT"
retweet: 3
to_user_id: null
to_user_id_str: null
}
```

Both Jaql and JSON are record-oriented models, and thus fit together perfectly. Note that JSON is not the only format that Jaql supports—in fact, Jaql is extremely flexible and can support many semistructured data sources such as XML, CSV, flat files, and more. However, in consideration of the space we have, we'll use the JSON example above in the following Jaql queries. As you will see from this section, Jaql looks very similar to Pig but also has some similarity to SQL.

Jaql Operators

Jaql is built on a set of core operators. Let's look at some of the most popular operators found in Jaql, how they work, and then go through some simple examples that will allow us to query the Twitter feed represented earlier.

FILTER The FILTER operator takes an array as input and filters out the elements of interest based on a specified predicate. For those familiar with SQL, think of the FILTER operator as a WHERE clause. For example, if you want to look only at the input records from the Twitter feed that were creat-ed by user `eatonchris`, you'd put something similar to the following in your query:

```
filter $.from_user == "eatonchris"
```

If you wanted to see only the tweets that have been retweeted more than twice, you would include a Jaql query such as this:

```
filter $.retweet > 2
```

TRANSFORM   The TRANSFORM operator takes an array as input and out-puts another array where the elements of the first array have been trans-formed in some way. For SQL addicts, you'll find this similar to the SELECT clause. For example, if an input array has two numbers denoted by N1 and N2, the TRANSFORM operator could produce the sum of these two numbers using the following:

```
transform { sum: $.N1 + $.N2 }
```

GROUP   The GROUP operator works much like the GROUP BY clause in SQL, where a set of data is aggregated for output. For example, if you wanted to count the total number of tweets in this section's working example, you could use this:

```
group into count($)
```

Likewise, if you wanted to determine the sum of all retweets by user, you would use a Jaql query such as this:

```
group by u = $.from_user into { total: sum($.retweet) };
```

JOIN   The JOIN operator takes two input arrays and produces an output array based on the join condition specified in the WHERE clause—similar to a join operation in SQL. Let's assume you have an array of tweets (such as the JSON tweet example) and you also have a set of interesting data that comes from a group of the people whom you follow on Twitter. Such an array may look like this:

```
following = { from_user: "eatonchris" },
            { from_user: "paulzikopoulos" }
```

In this example, you could use the JOIN operator to join the Twitter *feed* data with the Twitter *following* data to produce results for only the tweets from people you follow, like so:

```
join feed, follow
where feed.from_user = following.from_user
into {feed.*}
```

EXPAND   The EXPAND operator takes a nested array as input and produces a single array as output. Let's assume you have a nested array of geographic locations (denoted with latitude and longitude coordinates) as shown here:

```
geolocations = [[93.456, 123.222],[21.324, 90.456]]
```

In this case, the geolocations -> expand; command would return results in a single array as follows:

```
[93.456, 123.222, 21.324, 90.456]
```

SORT   As you might expect, the SORT operator takes an array as input and produces an array as output, where the elements are in a sorted order. The default Jaql sort order is ascending. You can sort Jaql results in a descending order using the sort by desc keyword.

TOP   The TOP operator returns the first n elements of the input array, where n is an <integer> that follows the TOP keyword.

Built-in Jaql Functions

In addition to the core operators, Jaql also has a large set of built-in functions that allow you to read in, manipulate, and write out data, as well as call external functions such as HDFS calls, and more. You can add your own custom-built functions, which can, in turn, invoke other functions. The more than 100 built-in functions are obviously too many to cover in this book; however, they are well documented in the base Jaql documentation.

A Jaql Query

Much like a MapReduce job is a flow of data, Jaql can be thought of as a pipeline of data flowing from a source, through a set of various operators, and out into a sink (a destination). The operand used to signify flow from one operand to another is an arrow: ->. Unlike SQL, where the output comes first (for example, the SELECT list), in Jaql, the operations listed are in natural order, where you specify the source, followed by the various operators you want to use to manipulate the data, and finally the sink.

Let's wrap up this Jaql section and put it all together with a simple Jaql example that counts the number of tweets written in English by user:

```
$tweets = read(hdfs("tweet_log"));
$tweets
-> filter $.iso_language_code = "en"
-> group by u = $.from_user
        into { user: $.from_user, total: sum($.retweet)
};
```

The first line simply opens up the file containing the data (with the intent to read it), which resides in HDFS, and assigns it a name, which in this case is $tweets. Next, the Jaql query reads $tweets and passes the data to the FILTER operator. The filter only passes on tweets that have an iso_language_code = en. These records are subsequently passed to the GROUP BY operator that adds the retweet values for each user together to get a sum for each given user.

Internally, the Jaql engine transforms the query into map and reduce tasks that can significantly reduce the application development time associated with analyzing massive amounts of data in Hadoop. Note that we've shown only the relationship between Jaql and JSON in this chapter; it's important to realize that this is not the only data format with which Jaql works. In fact, quite the contrary is true: Jaql is a flexible infrastructure for managing and analyzing many kinds of semistructured data such as XML, CSV data, flat files, relational data, and so on. In addition, from a development perspective, don't forget that the Jaql infrastructure is extremely flexible and extensible, and allows for the passing of data between the query interface and the application language of your choice (for example, Java, JavaScript, Python, Perl, Ruby, and so on).

Hadoop Streaming

In addition to Java, you can write map and reduce functions in other languages and invoke them using an API known as Hadoop Streaming (Streaming, for short). Streaming is based on the concept of UNIX streaming, where input is read from standard input, and output is written to standard output. These data streams represent the interface between Hadoop and your applications.

The Streaming interface lends itself best to short and simple applications you would typically develop using a scripting language such as Python or Ruby. A major reason for this is the text-based nature of the data flow, where each line of text represents a single record.

The following example shows the execution of map and reduce functions (written in Python) using Streaming:

For example:

```
hadoop  jar contrib/streaming/hadoop-streaming.jar \
    -input input/dataset.txt \
    -output output \
    -mapper text_processor_map.py \
    -reducer text_processor_reduce.py
```

Getting Your Data into Hadoop

One of the challenges with HDFS is that it's not a POSIX-compliant file system. This means that all the things you are accustomed to when it comes to interacting with a typical file system (copying, creating, moving, deleting, or accessing a file, and more) don't automatically apply to HDFS. To do anything with a file in HDFS, you must use the HDFS interfaces or APIs directly. That is yet another advantage of using the GPFS-SNC file system; with GPFS-SNC, you interact with your Big Data files in the same manner that you would any other file system, and, therefore, file manipulation tasks with Hadoop running on GPFS-SNC are greatly reduced. In this section, we discuss the basics of getting your data into HDFS and cover *Flume*, which is a distributed data collection service for flowing data into a Hadoop cluster.

Basic Copy Data

As you'll recall from the "Hadoop Common Components" section earlier in the chapter, you must use specific commands to move files into HDFS either through APIs or using the command shell. The most common way to move files from a local file system into HDFS is through the copyFromLocal command. To get files out of HDFS to the local file system, you'll typically use the `copyToLocal` command. An example of each of these commands is shown here:

```
hdfs dfs -copyFromLocal /user/dir/file hdfs://s1.n1.com/dir/hdfsfile
hdfs dfs -copyToLocal hdfs://s1.n1.com/dir/hdfsfile /user/dir/file
```

These commands are run through the HDFS shell program, which is simply a Java application. The shell uses the Java APIs for getting data into and out of HDFS. These APIs can be called from any Java application.

NOTE *HDFS commands can also be issued through the Hadoop shell, which is invoked by the command* `hadoop fs`.

The problem with this method is that you must have Java application developers write the logic and programs to read and write data from HDFS. Other methods are available (such as C++ APIs, or via the Thrift framework for cross-language services), but these are merely wrappers for the base Java APIs. If you need to access HDFS files from your Java applications, you would use the methods in the `org.apache.hadoop.fs` package. This allows you to incorporate read and write operations directly, to and from HDFS, from within your MapReduce applications. Note, however, that HDFS is designed for sequential read and write. This means when you write data to an HDFS file, you can write only to the end of the file (it's referred to as an `APPEND` in the database world). Herein lies yet another advantage to using GPFS-SNC as the file system backbone for your Hadoop cluster, because this specialized file system has the inherent ability to seek and write within a file, not just at the end of a file.

Flume

A flume is a channel that directs water from a source to some other location where water is needed. As its clever name implies, *Flume* was created (as of the time this book was published, it was an incubator Apache project) to allow you to flow data from a source into your Hadoop environment. In Flume, the entities you work with are called *sources, decorators,* and *sinks.* A *source* can be any data source, and Flume has many predefined source adapters, which we'll discuss in this section. A *sink* is the target of a specific operation (and in Flume, among other paradigms that use this term, the sink of one operation can be the source for the next downstream operation). A *decorator* is an operation on the stream that can transform the stream in some manner, which could be to compress or uncompress data, modify data by adding or removing pieces of information, and more.

A number of predefined source adapters are built into Flume. For example, some adapters allow the flow of anything coming off a TCP port to enter the flow, or anything coming to standard input (stdin). A number of text file source adapters give you the granular control to grab a specific file and feed it into a data flow or even take the tail of a file and continuously feed the flow with whatever new data is written to that file. The latter is very useful for feeding diagnostic or web logs into a data flow, since they are constantly being appended to, and the TAIL operator will continuously grab the latest entries from the file and put them into the flow. A number of other predefined source adapters, as well as a command exit, allow you to use any executable command to feed the flow of data.

There are three types of sinks in Flume. One sink is basically the final flow destination and is known as a *Collector Tier Event* sink. This is where you would land a flow (or possibly multiple flows joined together) into an HDFS-formatted file system. Another sink type used in Flume is called an *Agent Tier Event*; this sink is used when you want the sink to be the input source for another operation. When you use these sinks, Flume will also ensure the integrity of the flow by sending back acknowledgments that data has actually arrived at the sink. The final sink type is known as a *Basic* sink, which can be a text file, the console display, a simple HDFS path, or a null bucket where the data is simply deleted.

Look to Flume when you want to flow data from many sources (it was designed for log data, but it can be used for other kinds of data too), manipulate it, and then drop it into your Hadoop environment. Of course, when you want to perform very complex transformations and cleansing of your data, you should be looking at an enterprise-class data quality toolset such as IBM Information Server, which provides services for transformation, extraction, discovery, quality, remediation, and more. IBM Information Server can handle large-scale data manipulations prior to working on the data in a Hadoop cluster, and integration points are provided (with more coming) between the technologies (for instance the ability to see data lineage).

Other Hadoop Components

Many other open source projects fall under the Hadoop umbrella, either as Hadoop subprojects or as top-level Apache projects, with more popping up as time goes on (and as you may have guessed, their names are just as interesting: ZooKeeper, HBase, Oozie, Lucene, and more). In this section, we cover four additional Hadoop-related projects that you might encounter (all of which are shipped as part of any InfoSphere BigInsights edition).

ZooKeeper

ZooKeeper is an open source Apache project that provides a centralized infrastructure and services that enable synchronization across a cluster. ZooKeeper maintains common objects needed in large cluster environments. Examples of these objects include configuration information, hierarchical naming space, and so on. Applications can leverage these services to coordinate distributed processing across large clusters.

Imagine a Hadoop cluster spanning 500 or more commodity servers. If you've ever managed a database cluster with just 10 servers, you know there's a need for centralized management of the entire cluster in terms of name services, group services, synchronization services, configuration management, and more. In addition, many other open source projects that leverage Hadoop clusters require these types of cross-cluster services, and having them available in ZooKeeper means that each of these projects can embed ZooKeeper without having to build synchronization services from scratch into each project. Interaction with ZooKeeper occurs via Java or C interfaces at this time (our guess is that in the future the Open Source community will add additional development languages that interact with ZooKeeper).

ZooKeeper provides an infrastructure for cross-node synchronization and can be used by applications to ensure that tasks across the cluster are serialized or synchronized. It does this by maintaining status type information in memory on ZooKeeper servers. A ZooKeeper server is a machine that keeps a copy of the state of the entire system and persists this information in local log files. A very large Hadoop cluster can be surpported by multiple ZooKeeper servers (in this case, a master server synchronizes the top-level servers). Each client machine communicates with one of the ZooKeeper servers to retrieve and update its synchronization information.

Within ZooKeeper, an application can create what is called a *znode* (a file that persists in memory on the ZooKeeper servers). The znode can be updated by any node in the cluster, and any node in the cluster can register to be informed of changes to that znode (in ZooKeeper parlance, a server can be set up to "watch" a specific znode). Using this znode infrastructure (and there is much more to this such that we can't even begin to do it justice in this section), applications can synchronize their tasks across the distributed cluster by updating their status in a ZooKeeper znode, which would then inform the rest of the cluster of a specific node's status change. This cluster-wide status centralization service is essential for management and serialization tasks across a large distributed set of servers.

HBase

HBase is a column-oriented database management system that runs on top of HDFS. It is well suited for sparse data sets, which are common in many Big Data use cases. Unlike relational database systems, HBase does not support a structured query language like SQL; in fact, HBase isn't a relational data store at all. HBase applications are written in Java much like a typical MapReduce application. HBase does support writing applications in Avro, REST, and Thrift. (We briefly cover Avro at the end of this chapter, and the other two aren't covered in this book, but you can find details about them easily with a simple Google search.)

An HBase system comprises a set of tables. Each table contains rows and columns, much like a traditional database. Each table must have an element defined as a Primary Key, and all access attempts to HBase tables must use this Primary Key. An HBase column represents an attribute of an object; for example, if the table is storing diagnostic logs from servers in your environment, where each row might be a log record, a typical column in such a table would be the timestamp of when the log record was written, or perhaps the servername where the record originated. In fact, HBase allows for many attributes to be grouped together into what are known as *column families*, such that the elements of a column family are all stored together. This is different from a row-oriented relational database, where all the columns of a given row are stored together. With HBase you must predefine the table schema and specify the column families. However, it's very flexible in that new columns can be added to families at any time, making the schema flexible and therefore able to adapt to changing application requirements.

Just as HDFS has a NameNode and slave nodes, and MapReduce has JobTracker and TaskTracker slaves, HBase is built on similar concepts. In HBase a *master node* manages the cluster and *region servers* store portions of the tables and perform the work on the data. In the same way HDFS has some enterprise concerns due to the availability of the NameNode (among other areas that can be "hardened" for true enterprise deployments by BigInsights), HBase is also sensitive to the loss of its master node.

Oozie

As you have probably noticed in our discussion on MapReduce capabilities, many jobs might need to be chained together to satisfy a complex application. *Oozie* is an open source project that simplifies workflow and coordination between jobs. It provides users with the ability to define actions and dependencies between actions. Oozie will then schedule actions to execute when the required dependencies have been met.

A workflow in Oozie is defined in what is called a *Directed Acyclical Graph (DAG)*. *Acyclical* means there are no loops in the graph (in other words, there's a starting point and an ending point to the graph), and all tasks and dependencies point from start to end without going back. A DAG is made up of *action nodes* and *dependency nodes*. An action node can be a MapReduce job, a Pig application, a file system task, or a Java application. Flow control in the graph is represented by node elements that provide logic based on the input from the preceding task in the graph. Examples of flow control nodes are decisions, forks, and join nodes.

A workflow can be scheduled to begin based on a given time or based on the arrival of some specific data in the file system. After inception, further workflow actions are executed based on the completion of the previous actions in the graph. Figure 4-3 is an example of an Oozie workflow, where the nodes represent the actions and control flow operations.

Lucene

Lucene is an extremely popular open source Apache project for text search and is included in many open source projects. Lucene predates Hadoop and

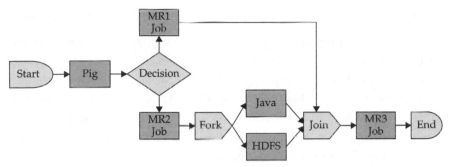

Figure 4-3 *An Oozie workflow that includes multiple decision points as part of the end-to-end execution*

has been a top-level Apache project since 2005. Lucene provides full text indexing and searching libraries for use within your Java application. (Note that Lucene has been ported to C++, Python, Perl, and more.) If you've searched on the Internet, it's likely that you've interacted with Lucene (although you probably didn't know it).

The Lucene concept is fairly simple, yet the use of these search libraries can be very powerful. In a nutshell, let's say you need to search within a collection of text, or a set of documents. Lucene breaks down these documents into text fields and builds an index on these fields. The index is the key component of Lucene, as it forms the basis for rapid text search capabilities. You then use the searching methods within the Lucene libraries to find the text components. This indexing and search platform is shipped with BigInsights and is integrated into Jaql, providing the ability to build, scan, and query Lucene indexes within Jaql.

BigInsights adds even greater capabilities by shipping a very robust text extraction library to glean structure out of unstructured text, which natively runs on BigInsights and leverages MapReduce. There's even a development framework to extend and customize the library with a complete tooling environment to make it relatively easy to use. By adding these text extractors to the text indexing capability, BigInsights provides one of the most feature-rich and powerful text analytics platforms for Hadoop available on the market today. What's more, you can't store a Lucene index in HDFS; however, you can store it with your other Hadoop data in GPFS-SNC.

Avro

Avro is an Apache project that provides data serialization services. When writing Avro data to a file, the schema that defines that data is always written to the file. This makes it easy for any application to read the data at a later time, because the schema defining the data is stored within the file. There's an added benefit to the Avro process: Data can be versioned by the fact that a schema change in an application can be easily handled because the schema for the older data remains stored within the data file. An Avro schema is defined using JSON, which we briefly discussed earlier in the "Jaql" section.

A schema defines the data types contained within a file and is validated as the data is written to the file using the Avro APIs. Similarly, the data can be formatted based on the schema definition as the data is read back from the file. The schema allows you to define two types of data. The first are the *primitive data types* such as STRING, INT[eger], LONG, FLOAT, DOUBLE, BYTE, NULL, and BOOLEAN. The second are *complex type definitions*. A complex type can be a record, an array, an enum (which defines an enumerated list of possible values for a type), a map, a union (which defines a type to be one of several types), or a fixed type.

APIs for Avro are available in C, C++, C#, Java, Python, Ruby, and PHP, making it available to most application development environments that are common around Hadoop.

Wrapping It Up

As you can see, Hadoop is more than just a single project, but rather an ecosystem of projects all targeted at simplifying, managing, coordinating, and analyzing large sets of data. IBM InfoSphere BigInsights fully embraces this ecosystem with code committers, contributions, and a no-fork backwards compatibility commitment. In the next chapter, we'll specifically look at the things IBM does to extend Hadoop and its related technologies into an analytics platform enriched with the enterprise-class experience IBM brings to this partnership.

5

InfoSphere BigInsights:
Analytics for Big
Data at Rest

Hadoop offers a great deal of potential in enabling enterprises to harness the data that was, until now, difficult to manage and analyze. Specifically, Hadoop makes it possible to process extremely large volumes of data with varying structures (or no structure at all). That said, with all the promise of Hadoop, it's still a relatively young technology. The Apache Hadoop top-level project was started in 2006, and although adoption rates are increasing— along with the number of open source code contributors—Hadoop still has a number of known shortcomings (in fairness, it's not even at version 1.0 yet). From an enterprise perspective, these shortcomings could either prevent companies from using Hadoop in a production setting, or impede its adoption, because certain operational qualities are always expected in production, such as performance, administrative capabilities, and robustness. For example, as we discussed in Chapter 4, the Hadoop Distributed File System (HDFS) has a centralized metadata store (referred to as the *NameNode*), which represents a single point of failure (SPOF) without availability (a cold standby is added in version 0.21). When the NameNode is recovered, it can take a long time to get the Hadoop cluster running again, because the metadata it tracks has to be loaded into the NameNode's memory structures, which all need to be rebuilt and repopulated. In addition, Hadoop can be complex to install, configure, and administer, and there isn't yet a significant number of people with Hadoop skills. Similarly, there is a limited pool of developers who have MapReduce skills. Programming traditional analytic algorithms (such as statistical or text analysis) to work in Hadoop is difficult, and it requires analysts to become skilled Java programmers with the ability to apply MapReduce techniques to their analytics algorithms. (Higher level languages, such as Pig and Jaql, make MapReduce programming easier, but still have a learning curve.) There's more, but you get the point: not only is Hadoop in need of some enterprise hardening, but also tools and features that help round out the possibilities of what the Hadoop platform offers (for example, visualization, text analytics, and graphical administration tools).

IBM InfoSphere BigInsights (BigInsights) addresses all of these issues, and more, through IBM's focus on two primary product goals:

- Deliver a Hadoop platform that's *hardened* for enterprise use with deep consideration for high availability, scalability, performance, ease-of-use, and other things you've come to expect from any solution you'll deploy in your enterprise.

- Flatten the time-to-value curve associated with Big Data analytics by providing the development and runtime environments for developers to build advanced analytical applications, and providing tools for business users to analyze Big Data.

In this chapter, we'll talk about how IBM readies Hadoop for enterprise usage. As you can imagine, IBM has a long history of understanding enterprise needs. By embracing new technologies such as Hadoop (and its open source ecosystem) and extending them with the deep experience and intellectual capital that IBM has built over its century of existence, you gain a winning combination that lets you explore Hadoop with a platform you can trust that also yields results more quickly.

Ease of Use: A Simple Installation Process

The BigInsights installer was designed with simplicity in mind. IBM's development teams asked themselves, "How can IBM cut the time-to-Hadoop curve without the effort and technical skills normally required to get open source software up and running?" They answered this question with the BigInsights installer.

The main objective of the BigInsights installer is to insulate you from complexity. As such, you don't need to worry about software prerequisites or determining which Apache Hadoop components to download, the configuration between these components, and the overall setup of the Hadoop cluster. The BigInsights installer does all of this for you, and all you need to do is click a button. Hadoop startup complexity is all but eliminated with BigInsights. Quite simply, your experience is going to be much like installing any commercial software.

To prepare for the writing of this book, we created three different Hadoop clusters:

- One from scratch, using just open source software, which we call the roll your own (RYO) Hadoop approach

- One from a competitor who solely offers an installation program, some operational tooling, and a Hadoop support contract

- One with BigInsights

The "roll your own" Hadoop approach had us going directly to the Apache web site and downloading the Hadoop projects, which ended up involving a lot of work. Specifically, we had to do the following:

1. Choose which Hadoop components, and which versions of those components, to download. We found many components, and it wasn't immediately obvious to us which ones we needed in order to start our implementation, so that required some preliminary research.

2. Create and set up a Hadoop user account.

3. Download each of the Hadoop components we decided we needed and install them on our cluster of machines.

4. Configure Secure Shell (SSH) for the Hadoop user account, and copy the keys to each machine in the cluster.

5. Configure Hadoop to define how we wanted it to run; for example, we specified I/O settings, JobTracker, and TaskTracker-level details.

6. Configure HDFS—specifically, we set up and formatted the NameNode and Secondary NameNode.

7. Define all of the global variables (for example, HADOOP_CLASSPATH, HADOOP_PID_DIR, HADOOP_HEAPSIZE, JAVA_HOME).

As you can imagine, getting the Hadoop cluster up and running from the open source components was complex and somewhat laborious. We got through it, but with some effort. Then again, we have an army of experienced Hadoop developers ready to answer questions. If you're taking the RYO route, you'll need a good understanding of the whole Hadoop ecosystem, along with basic Hadoop administration and configuration skills. You need to do all this before you can even begin to think about running simple MapReduce jobs, let alone running any kind of meaningful analytics applications.

Next, we tried installing a competitor's Hadoop distribution (notice it's a distribution, not a platform like BigInsights). This competitor's installation did indeed represent an improvement over the no-frills open source approach because it has a nifty graphical installer. However, it doesn't install and configure additional Hadoop ecosystem components such as Pig, Hive, and Flume, among others, which you need to install manually.

These two experiences stand in contrast to the BigInsights approach of simply laying down and configuring the entire set of required components using a single installer. With BigInsights, installation requires only a few button clicks, eliminating any worry about all the Hadoop-related components and versions. Very little configuration is needed and no extra prerequisites need to be downloaded. What's more, you can use IBM's installation program to graphically build a response file, which you can subsequently use to deploy BigInsights on all the nodes in your cluster in an automated fashion.

Hadoop Components Included in BigInsights 1.2

BigInsights features Apache Hadoop and its related open source projects as a core component. IBM remains committed to the integrity of the open source projects, and will not fork or otherwise deviate from their code. The following table lists the open source projects (and their versions) included in BigInsights 1.2, which was the most current version available at the time of writing:

Component	Version
Hadoop (common utilities, HDFS, and the MapReduce framework)	0.20.2
Jaql (programming and query language)	0.5.2
Pig (programming and query language)	0.7
Flume (data collection and aggregation)	0.9.1

Component	Version
Hive (data summarization and querying)	0.5
Lucene (text search)	3.1.0
ZooKeeper (process coordination)	3.2.2
Avro (data serialization)	1.5.1
HBase (real-time read and write database)	0.20.6
Oozie (workflow and job orchestration)	2.2.2

With each release of BigInsights, updates to the open source components and IBM components go through a series of testing cycles to ensure that they work together. That's another pretty special point we want to clarify: You can't just drop new code into production. Some backward-compatibility issues are always present in our experience with open source projects. BigInsights takes away all of the risk and guesswork for your Hadoop components. It goes through the same rigorous regression and quality assurance testing processes used for other IBM software. So ask yourself this: Would you rather be your own systems integrator, testing all of the Hadoop components repeatedly to ensure compatibility? Or would you rather let IBM find a stable stack that you can deploy and be assured of a reliable working environment?

Finally, the BigInsights installer also lays down additional infrastructure, including analytics tooling and components that provide enterprise stability and quality to Hadoop, which is what makes BigInsights a platform instead of a distribution. We'll discuss those in the remainder of this chapter.

A Hadoop-Ready Enterprise-Quality File System: GPFS-SNC

The General Parallel File System (GPFS) was developed by IBM Research in the 1990s for High-Performance Computing (HPC) applications. Since its first release in 1998, GPFS has been used in many of the world's fastest super-computers, including Blue Gene, Watson (the *Jeopardy!* supercomputer), and ASC Purple. (The GPFS installation in the ASC Purple supercomputer supported data throughput at a staggering 120 GB per second!) In addition to HPC, GPFS is commonly found in thousands of other mission-critical installations worldwide. GPFS is also the file system that's part of DB2 pureScale and is even found standing up many Oracle RAC installations; you'll also find GPFS underpinning highly scalable web and file servers, other data-bases, applications in the finance and engineering sectors, and more. Need-less to say, GPFS has earned an enterprise-grade reputation and pedigree for extreme scalability, high performance, and reliability.

One barrier to the widespread adoption of Hadoop in some enterprises to-day is HDFS. It's a relatively new file system with some design-oriented limi-tations. The principles guiding the development of HDFS were defined by use cases, which assumed Hadoop workloads would involve sequential reads of very large file sets (and no random writes to files already in the cluster—just append writes). In contrast, GPFS has been designed for a wide variety of workloads and for a multitude of uses, which we'll talk about in this section.

Extending GPFS for Hadoop: GPFS Shared Nothing Cluster

GPFS was originally available only as a storage area network (SAN) file sys-tem, which isn't suitable for a Hadoop cluster since these clusters use locally attached disks. The reason why SAN technology isn't optimal for Hadoop is because MapReduce jobs perform better when data is stored on the node where it's processed (which requires locality awareness for the data). In a SAN, the location of the data is transparent, which results in a high degree of network bandwidth and disk I/O, especially in clusters with many nodes.

In 2009, IBM extended GPFS to work with Hadoop with *GPFS-SNC* (*Shared Nothing Cluster*). The following are the key additions IBM made to GPFS that allows it to be a suitable file system for Hadoop, thereby hardening Hadoop for the enterprise:

- **Locality awareness**   A key feature of Hadoop is that it strives to process data at the node where the data is stored. This minimizes network traffic and improves performance. To support this, GPFS-SNC provides location information for all files stored in the cluster. The Hadoop JobTracker uses this location information to pick a replica that is local to the task that needs to be run, which helps increase performance.

- **Meta-blocks**   The typical GPFS block size is 256 KB, while in a Hadoop cluster, the block size is much larger. For instance, recommended block size for BigInsights is 128 MB. In GPFS-SNC, a collection of many standard GPFS blocks are put together to create the concept of a *meta-block*. Individual map tasks execute against meta-blocks, while file operations outside Hadoop will still use the normal smaller block size, which is more efficient for other kinds of applications. This flexibility enables a variety of applications to work on the same cluster, while maintaining optimal performance. HDFS doesn't share these benefits, as its storage is restricted to Hadoop and Hadoop alone. For example, you can't host a Lucene text index on HDFS. In GPFS-SNC, however, you can store a Lucene text index alongside the text data in your cluster (this co-location has performance benefits). Although Lucene uses the GPFS block size of 256 KB for its operations, any Hadoop data is stored in the cluster and read in meta-blocks.

- **Write affinity and configurable replication**   GPFS-SNC allows you to define a placement strategy for your files, including the approach taken during file replication. The normal replication policy is for the first copy to be local, the second copy to be local to the rack (which is different than HDFS), and the third copy to be striped across other racks in the cluster. GPFS-SNC lets you customize this replication strategy. For instance, you might decide that a specific set of files should always be stored together to allow an application to access the data from the same location. This is something you can't do in HDFS, which can lead to higher performance for specific workloads such as large sequential reads. The strategy for the second replica could also be to keep this same data together. In case the primary node fails, it would be easy to switch to another node without seeing any degradation in application performance. The third copy of the data is typically stored striped, in case one of the first two copies must be rebuilt. Restoring files is much faster to do when files are striped. In HDFS, there is no data striping, and you cannot customize write affinity or replication behavior (except to change the replication factor).

- **Configurable recovery policy**   When a disk fails, any files with lost blocks become under-replicated. GPFS-SNC will automatically copy the missing files in the cluster to maintain the replication level. GPFS-SNC lets you customize the policy of what to do in the event of a disk failure. For example, one approach could be to restripe the disk when a failure occurs. Since one of the replicated copies of a file is typically striped, the rebuilding of missing blocks is very fast as reads are done in parallel.

Alternatively, you could specify a policy to rebuild the disk—perhaps, for example, when the disk is replaced. These recovery policies don't need to be automatic; you could decide to use manual recovery in the case of a maintenance task, such as swapping a set of disks or nodes. You could also configure recovery to work incrementally. For example, if a disk was offline and was brought online later, GPFS-SNC knows to copy only its missing blocks, as it maintains the list of files on each disk. In HDFS, the NameNode will initiate replication for files that are under-replicated, but recovery is not customizable.

All of the characteristics that make GPFS the file system of choice for large-scale mission-critical IT installations are applicable to GPFS-SNC. After all, this is still GPFS, but with the Hadoop-friendly extensions. You get the same stability, flexibility, and performance in GPFS-SNC, as well as all of the utilities that you're used to. GPFS-SNC also provides hierarchical storage management (HSM) capabilities, where it can manage and use disk drives with different retrieval speeds efficiently. This enables you to manage multi-temperature data, keeping your hot data on your best performing hardware. HDFS doesn't have this ability.

GPFS-SNC is such a game changer that it won the prestigious Supercomputing Storage Challenge award in 2010 for being the "most innovative storage solution" submitted to this competition.

What Does a GPFS-SNC Cluster Look Like?

GPFS-SNC is a distributed storage cluster with a shared-nothing architecture. There is no central store for metadata because it's shared across multiple nodes in the cluster. Additionally, file system management tasks are distributed between the data nodes in the cluster, such that if a failure occurs, replacement nodes are nominated to assume these tasks automatically.

As you can see in Figure 5-1, a GPFS-SNC cluster consists of multiple racks of commodity hardware, where the storage is attached to the compute nodes. If you're familiar with HDFS, you'll notice that Figure 5-1 doesn't include a NameNode, a Secondary NameNode, or anything that acts as a centralized metadata store. *This is a significant benefit of GPFS-SNC over HDFS.*

The cluster design in Figure 5-1 is a simple example and assumes that each compute node has the same CPU, RAM, and storage specifications (in reality there might be hardware differences with the Quorum Nodes and Primary

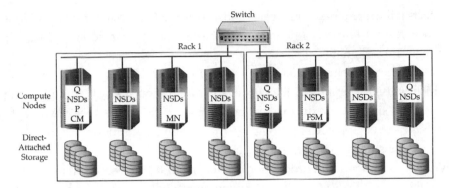

Figure 5-1   *An example of a GPFS-SNC cluster*

Cluster Configuration Server to harden them so that they are less likely to undergo an outage). To ensure smooth management of the cluster, different compute nodes in a GPFS-SNC cluster assume different management roles, which deserve some discussion.

The figure shows that each compute node in a GPFS-SNC cluster has a *Network Shared Disk (NSD)* server service that can access the local disks. When one node in a GPFS-SNC cluster needs to access data on a different node, the request goes through the NSD server. As such, NSD servers help move data between the nodes in the cluster.

A *Quorum Node (Q)* works together with other Quorum Nodes in a GPFS-SNC cluster to determine whether the cluster is running and available for incoming client requests. A Quorum Node is also used to ensure data consistency across a cluster in the event of a node failure. A cluster administrator designates the Quorum Node service to a selected set of nodes during cluster creation or while adding nodes to the cluster. Typically, you'll find one Quorum Node per rack, with the maximum recommended number of Quorum Nodes being seven. When setting up a GPFS-SNC cluster, an administrator should define an odd number of nodes and, if you don't have a homogeneous cluster, assign the Quorum Node role to a machine that is least likely to fail. If one of the Quorum Nodes is lost, the remaining Quorum Nodes will talk to each other to verify that quorum is still intact.

There's a single *Cluster Manager (CM)* node per GPFS-SNC cluster, and it's selected by the Quorum Nodes (as opposed to being designated by the cluster administrator). The Cluster Manager determines quorum, manages disk leases, detects failures, manages recovery, and selects the File System Manager Node. If the Cluster Manager node fails, the Quorum Nodes immediately detect the outage and designate a replacement.

A GPFS-SNC cluster also has a *Primary Cluster Configuration Server (P)*, which is used to maintain the cluster's configuration files (this role is designated to a single node during cluster creation). If this node goes down, automated recovery protocols are engaged to designate another node to assume this responsibility. The *Secondary Configuration Server (S)* is optional, but we highly recommend that production clusters include one because it removes an SPOF in the GPFS-SNC cluster by taking over the Primary Cluster Configuration Server role in the event of a failure. If the Primary Cluster Configuration Server and the Secondary Cluster Configuration Server both fail, the cluster configuration data will still be intact (since cluster configuration data is replicated on all nodes), but manual intervention will be required to revive the cluster.

Each GPFS-SNC cluster has one or more *File System Manager (FSM)* nodes that are chosen dynamically by the Cluster Manager node (although a cluster administrator can define a pool of available nodes for this role). The File System Manager is responsible for file system configuration, usage, disk space allocation, and quota management. This node can have higher memory and CPU demands than other nodes in the cluster; generally, we recommend that larger GPFS-SNC clusters have multiple File System Managers.

The final service in Figure 5-1 is the *Metanode (MN)*. There's a Metanode for each open file in a GPFS-SNC cluster, and it's responsible for maintaining file metadata integrity. In almost all cases, the Metanode service will run on the node where the particular file was open for the longest continuous period of time. All nodes accessing a file can read and write data directly, but updates to metadata are written only by the Metanode. The Metanode for each file is independent of that for any other file, and can move to any node to meet application requirements.

A *Failure Group* is defined as a set of disks that share a common point of failure that could cause them all to become unavailable at the same time. For example, all the disks in an individual node in a cluster form a failure group, because if this node fails, all disks in the node immediately become unavailable. The GPFS-SNC approach to replication reflects the concept of a Failure Group, as the cluster will ensure that there is a copy of each block of replicated data and metadata on disks belonging to different failure groups. Should a set of disks become unavailable, GPFS-SNC will retrieve the data from the other replicated locations.

If you select the GPFS-SNC component for installation, the BigInsights graphical installer will handle the creation and configuration of the GPFS-SNC cluster for you. The installer prompts you for input on the nodes where it should assign the Cluster Manager and Quorum Node services. This installation approach uses default configurations that are typical for BigInsights workloads. GPFS-SNC is highly customizable, so for specialized installations, you can install and configure it outside of the graphical installer by modifying the template scripts and configuration files (although some customization is available within the installer itself).

GPFS-SNC Failover Scenarios

Regardless of whether you're using GPFS-SNC or HDFS for your cluster, the Hadoop MapReduce framework is running on top of the file system layer. A running Hadoop cluster depends on the TaskTracker and JobTracker services, which run on the GPFS-SNC or HDFS storage layers to support MapReduce workloads. Although these servers are not specific to the file system layer, they do represent an SPOF in a Hadoop cluster. This is because if the JobTracker node fails, all executing jobs fail as well; however, this kind of failure is rare and is easily recoverable. A NameNode failure in HDFS is far more serious and has the potential to result in data loss if its disks are corrupted and not backed up. In addition, for clusters with many terabytes of storage, restarting the NameNode can take hours, as the cluster's metadata needs to be fetched from disk and read into memory, and all the changes from the previous checkpoint must be replayed. In the case of GPFS-SNC, there is no need for a NameNode (it is solely an HDFS component).

Different kinds of failures can occur in a cluster, and we describe how GPFS-SNC handles each of these failure scenarios:

- **Cluster Manager Failure**   When the Cluster Manager fails, Quorum Nodes detect this condition and will elect a new Cluster Manager (from the pool of Quorum Nodes). Cluster operations will continue with a very small interruption in overall cluster performance.

- **File System Manager Node Failure**   Quorum Nodes detect this condition and ask the Cluster Manager to pick a new File System Manager Node from among any of the nodes in the cluster. This kind of failure results in a very small interruption in overall cluster performance.

- **Secondary Cluster Configuration Server Failure**   Quorum Nodes will detect the failure, but the cluster administrator will be required to designate a new node manually as the Secondary Cluster Configuration Server. Cluster operations will continue even if this node is in a failure state, but some administrative commands that require both the primary and secondary servers might not work.

- **Rack Failure**   The remaining Quorum Nodes will decide which part of the cluster is still operational and which nodes went down with it. If the Cluster Manager was on the rack that went down, Quorum Nodes will elect a new Cluster Manager in the healthy part of the cluster. Similarly, the Cluster Manager will pick a File System Manager Node in case the old one was on the failed rack. The cluster will employ standard recovery strategies for each of the individual data nodes lost on the failed rack.

GPFS-SNC POSIX-Compliance

A significant architectural difference between GPFS-SNC and HDFS is that *GPFS-SNC is a kernel-level file system, while HDFS runs on top of the operating system*. As a result, HDFS inherently has a number of restrictions and inefficiencies. Most of these limitations stem from the fact that HDFS is not fully POSIX-compliant. On the other hand, GPFS-SNC is 100 percent POSIX-compliant. This makes your Hadoop cluster more stable, more secure, and more flexible.

Ease of Use and Storage Flexibility

Files stored in GPFS-SNC are visible to all applications, just like any other files stored on a computer. For instance, when copying files, any authorized user can use traditional operating system commands to list, copy, and move files in GPFS-SNC. This isn't the case in HDFS, where users need to log into Hadoop to see the files in the cluster. In addition, if you want to perform any file manipulations in HDFS, you need to understand how the Hadoop command shell environment works and know specific Hadoop file system commands.

All of this results in extra training for IT staff. Experienced administrators can get used to it, but it's a learning curve nevertheless. As for replication or backups, the only mechanism available for HDFS is to copy files manually through the Hadoop command shell.

The full POSIX compliance of BigInsights' GPFS-SNC enables you to manage your Hadoop storage just as you would any other computer in your IT environment. That's going to give you economies of scale when it comes to building Hadoop skills and just making life easier. For example, your traditional file administration utilities will work, as will your backup and restore tooling and procedures. GPFS-SNC will actually extend your backup capabilities as it includes point-in-time (PiT) snapshot backup, off-site replication, and other utilities.

With GPFS-SNC, other applications can even share the same storage resources with Hadoop. This is not possible in HDFS, where you need to define disk space dedicated to the Hadoop cluster up front. Not only must you estimate how much data you need to store in HDFS, but you must also guess how much storage you'll need for the output of MapReduce jobs, which can vary widely by workload; don't forget you need to account for space that will be taken up by log files created by the Hadoop system too! With GPFS-SNC, you only need to worry about the disks themselves filling up; there is no need to dedicate storage for Hadoop.

Concurrent Read/Write

An added benefit of GPFS-SNC's POSIX compliance is that it gives your MapReduce applications—or any other application, for that matter—the ability to update existing files in the cluster without simply appending to them. In addition, GPFS-SNC enables multiple applications to concurrently write to the same file in the Hadoop cluster. Again, neither of these capabilities is possible with HDFS, and these file-write restrictions limit what HDFS can do for your Big Data ecosystem. For example, BigIndex, or a Lucene text index (which is an important component for any kind of meaningful text processing and analytics), can readily be used in GPFS-SNC. As we discussed before, you are using HDFS, Lucene needs to maintain its indexes in the local file system, and not in HDFS, because Lucene needs to update existing files continually, and all of this (you guessed it) adds complexity and performance overhead.

Security

As previously mentioned, unlike HDFS, GPFS-SNC is a kernel-level file system, which means it can take advantage of operating system–level security. Extended permissions through POSIX Access Control Lists (ACLs) are also possible, which enables fine-grained user-specific permissions that just aren't possible in HDFS.

With the current Hadoop release (0.20 as of the time this book was written), HDFS is not aware of operating system–level security, which means anyone with access to the cluster can read its data. Although Hadoop 0.21 and 0.22 will integrate security capabilities into HDFS (which will require users to be authenticated and authorized to use the cluster), this new security model is more complex to administer and is less flexible than what's offered in GPFS-SNC. (We talk more about security later in this chapter.)

GPFS-SNC Performance

The initial purpose for GPFS was to be a storage system for high-performance supercomputers. With this high-performance lineage, GPFS-SNC has three key features that give it the flexibility and power that enable it to consistently outperform HDFS.

The first feature is *data striping*. In GPFS-SNC, the cluster stripes and mirrors everything (SAME) so data is striped across the disks in the cluster. The striping enables sequential reads to be faster than when performed on HDFS because this data can be read and processed in parallel. This greatly aids operations such as sorts, which require high sequential throughputs. In HDFS, the files are replicated across the cluster according to the replication factor, but there is no striping of individual blocks across multiple disks.

Another performance booster for GPFS-SNC is *distributed metadata*. In GPFS-SNC, file metadata is distributed across the cluster, which improves the performance of workloads with many random block reads. In HDFS, metadata is stored centrally on the NameNode, which is not only a single point of failure, but also a performance bottleneck for random access workloads.

Random access workloads in a GPFS-SNC cluster get an additional performance boost because of *client-side caching*. There's just no caching like this in HDFS. Good random access performance is important for Hadoop workloads, in spite of the underlying design, which favors sequential access. For example,

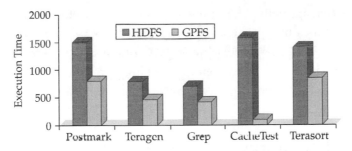

Figure 5-2   *Performance benchmark comparing GPFS to HDFS*

exploratory analysis activities with Pig and Jaql applications benefit greatly from good random I/O performance.

IBM Research performed benchmark testing of GPFS-SNC and HDFS using standard Hadoop workloads on the same clusters (one with GPFS-SNC and one running HDFS). The impressive performance boost (and our lawyers wouldn't let us go to print without this standard disclaimer: your results may vary) is shown in Figure 5-2.

As you can see, GPFS-SNC gives a significant performance boost for Hadoop workloads, when compared to the default HDFS file system. With these results, and within our own internal tests, we estimated that a 10-node Hadoop cluster running on GPFS-SNC will perform at the same level as a Hadoop cluster running on HDFS with approximately 16 of the same nodes.

GPFS-SNC Hadoop Gives Enterprise Qualities

We've spent a lot of time detailing all of the benefits that GPFS-SNC provides a Hadoop cluster because of their importance. These benefits showcase how IBM's assets, experiences, and research can harden and simultaneously complement the innovations from the Hadoop open source community, thereby creating the foundation for an enterprise-grade Big Data platform. In summary, there are availability, security, performance, and manageability advantages to leveraging GPFS-SNC in your Hadoop cluster.

Compression

When dealing with the large volumes of data expected in a Hadoop setting, the idea of compression is appealing. On the one hand, you can save a great deal of space (especially when considering that every storage block is replicated three times by default in Hadoop); on the other hand, data transfer speeds are improved because of lower data volumes on the wire. You should consider two important items before choosing a compression scheme: *splittable compression* and the *compression and decompression speeds* of the compression algorithm you're using.

Splittable Compression

In Hadoop, files are split (divided) if they are larger than the cluster's block size setting (normally one split for each block). For uncompressed files, this means individual file splits can be processed in parallel by different mappers. Figure 5-3 shows an uncompressed file with the vertical lines representing the split and block boundaries (in this case, the split and block size are the same).

When files, especially text files, are compressed, complications arise. For most compression algorithms, individual file splits cannot be decompressed independently from the other splits from the same file. More specifically, these compression algorithms are not *splittable* (remember this key term when discussing compression and Hadoop). In the current release of Hadoop (0.20.2 at the time of writing), no support is provided for splitting compressed text files. For files in which the Sequence or Avro formats are applied, this is not an issue, because these formats have built-in synchronization points, and are therefore splittable. For unsplittable compressed text files, MapReduce processing is limited to a single mapper.

For example, suppose the file in Figure 5-3 is a 1 GB text file in your Hadoop cluster, and your block size is set at the BigInsights default of 128 MB, which means your file spans eight blocks. When this file is compressed using the conventional algorithms available in Hadoop, it's no longer possible to parallelize the processing for each of the compressed file splits, because the file can be decompressed only as a whole, and not as individual parts based on the splits. Figure 5-4 depicts this file in a compressed (and binary) state,

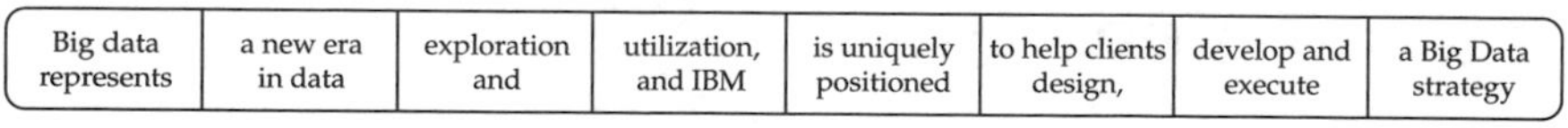

Big data represents	a new era in data	exploration and	utilization, and IBM	is uniquely positioned	to help clients design,	develop and execute	a Big Data strategy

Figure 5-3 *An uncompressed splittable file in Hadoop*

Figure 5-4   *A compressed nonsplittable file*

with the splits being impossible to decompress individually. Note that the split boundaries are dotted lines, and the block boundaries are solid lines.

Because Hadoop 0.20.2 doesn't support splittable text compression natively, all the splits for a compressed text file will be processed by only a single mapper. For many workloads, this would cause such a significant performance hit that it wouldn't be a viable option. However, Jaql is configured to understand splittable compression for text files and will process them automatically with parallel mappers. You can do this manually for other environments (such as Pig and MapReduce programs) by using the `TextInput-Format` input format instead of the Hadoop standard.

Compression and Decompression

The old saying "nothing in this world is free" is surely true when it comes to compression. There's no magic going on; in essence, you are simply consuming CPU cycles to save disk space. So let's start with this assumption: There could be a performance penalty for compressing data in your Hadoop cluster, because when data is written to the cluster, the compression algorithms (which are CPU-intensive) need CPU cycles and time to compress the data. Likewise, when reading data, any MapReduce workloads against compressed data can incur a performance penalty because of the CPU cycles and the time required to decompress the compressed data. This creates a conundrum: You need to balance priorities between storage savings and additional performance overhead.

NOTE   *If you've got an application that's I/O bound (typical for many warehouse-style applications), you might see a performance gain in your application, because I/O-bound systems typically have spare CPU cycles (found as idle I/O wait in the CPU) that can be utilized to run the compression and decompression algorithms. For example, if you use idle I/O wait CPU cycles to do the compression, and you get good compression rates, you could end up with more data flowing through the I/O pipe, and that means faster performance for those applications that need to fetch a lot of data from disk.*

A BigInsights Bonus: IBM LZO Compression

BigInsights includes the IBM LZO compression codec, which supports splitting compressed files and enabling individual compressed splits to be processed in parallel by your MapReduce jobs.

Some Hadoop online forums describe how to use the GNU version of LZO to enable splittable compression, so why did IBM create a version of it, and why not use the GNU LZO alternative? First, the IBM LZO compression codec *does not* create an index while compressing a file, because it uses fixed-length compression blocks. In contrast, the GNU LZO algorithm uses variable-length compression blocks, which leads to the added complexity of needing an index file that tells the mapper where it can safely split a compressed file. (For GNU LZO compression, this means mappers would need to perform index lookups during decompress and read operations. With this index, there is administrative overhead, because if you move the compressed file, you will need to move the corresponding index file as well.) Second, many companies, including IBM, have legal policies that prevent them from purchasing or releasing software that includes GNU Public License (GPL) components. This means that the approach described in online Hadoop forums requires additional administrative overhead and configuration work. In addition, there are businesses with policies restricting the deployment of GPL code. The IBM LZO compression is fully integrated with BigInsights and under the *same* enterprise-friendly license agreement as the rest of BigInsights, which means you can use it with less hassle and none of the complications associated with the GPL alternative.

In the next release of Hadoop (version 0.21), the bzip2 algorithm will support splitting. However, decompression speed for bzip2 is much slower than for IBM LZO, so bzip2 is not a desirable compression algorithm for workloads where performance is important.

Figure 5-5 shows the compressed text file from the earlier examples, but in a splittable state, where individual splits can be decompressed by their own

| 0001 1010 | 0001 1101 | 1100 0100 | 1010 1110 | 0101 1100 | 1101 0011 | 0001 1010 | 1101 1100 |

Figure 5-5   *A splittable compressed text file*

mappers. Note that the split sizes are equal, indicating the fixed-length compression blocks.

Compression Codec	File Extension	Splittable	Degree of Compression	Decompression Speed
IBM LZO	.cmx	Yes	Medium	Fastest
bzip2	.bz2	Yes, but not available until Hadoop 0.21	Highest	Slow
gzip	.gz	No	High	Fast
DEFLATE	.deflate	No	High	Fast

In the previous table you can see the four compression algorithms available on the BigInsights platform (IBM LZO, bzip2, gzip, and DEFLATE) and some of their characteristics.

Finally, the following table shows some benchmark comparison results for the three most popular compression algorithms commonly used in Hadoop (original source: http://stephane.lesimple.fr/wiki/blog/lzop_vs_compress_vs_gzip_vs_bzip2_vs_lzma_vs_lzma2-xz_benchmark_reloaded). In this benchmark, a 96 MB file is used as the test case. Note that the performance and compression ratio for the IBM LZO algorithm is on par with the LZO algorithm tested in this benchmark, but with the benefit of being splittable without having to use indexes, and being released under an enterprise-friendly license.

Codec	Compressed Size (MB)	Compression Speed (s)	Decompression Speed (s)
LZO	36	1	0.6
bzip2	19	22	5
gzip	23	10	1.3

Administrative Tooling

To aid in the administration of your cluster, BigInsights includes a web-based administration console that provides a real-time, interactive view of your cluster. The BigInsights console provides a graphical tool for examining the health of your BigInsights environment, including the nodes in your cluster, the status of your jobs (applications), and the contents of your HDFS or GPFS-SNC file system. It's automatically included in the BigInsights installation and by default runs on port 8080, although you can specify a different port during the installation process.

Apache Hadoop is composed of many disparate components, and each has its own configuration and administration considerations. In addition, Hadoop clusters are often large and impose a variety of administration challenges. The BigInsights administration console provides a single, harmonized view of your cluster that simplifies your work. Through this console, you can add and remove nodes, start and stop nodes, assess an application's status, inspect the status of MapReduce jobs, review log records, assess the overall health of your platform (storage, nodes, and servers), start and stop optional components (for example, ZooKeeper), navigate files in the BigInsights cluster, and more.

Figure 5-6 shows a snippet of the console's main page. As you can see, the administration console focuses on tasks required for administering your Hadoop cluster. A dashboard summarizes the health of your system, and you can drill down to get details about individual components.

In Figure 5-6, you can also see a tab for HDFS, which allows you to navigate through an HDFS directory structure to see what files have been stored and create new directories. You can also upload files to HDFS through this tool, although it's not well suited for large files. For uploading large files to your Hadoop cluster, we recommend using other mechanisms, such as Flume.

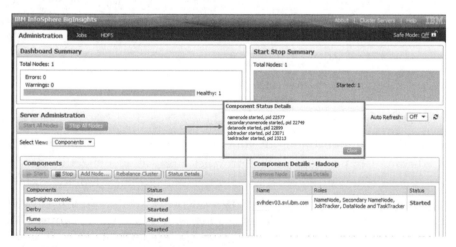

Figure 5-6   *An example of the BigInsights administration console*

The BigInsights console features a GPFS tab (if you are using GPFS-SNC as opposed to HDFS) which provides the same capability to browse and exchange data with the GPFS-SNC file system.

The main page of this web console also allows you to link to the Cluster Server tools provided by the underlying open source components. This comes in handy for administrators, because it easily allows them to access a variety of built-in tools from a single console.

Figure 5-7 shows a couple of screens from the administration console. On the top is the Job Status page, where you can view summary information for your cluster, such as status, the number of nodes that make up the cluster, task capacity, jobs in progress, and so on. If a job is in progress, and you have the appropriate authorizations, you can cancel the running job as shown at the bottom of this figure. To view details about a specific job, select a job from the list and view its details in the Job Summary section of the page. You can drill down even further to get more granular details about your jobs. For example, you can view the job configuration (shown as XML) and counter information, which explains how many mappers were used at execution time and the number of bytes read from/written to complete the job.

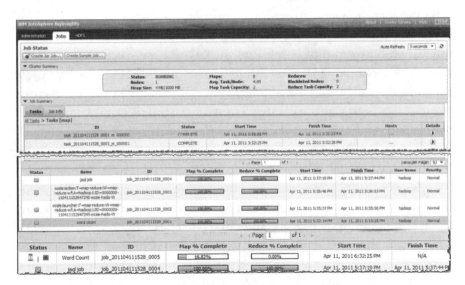

Figure 5-7   *Job Status and Jobs in Progress windows in the BigInsights administration console*

A number of other tooling benefits are provided in BigInsights that aren't available in a regular Hadoop environment. For example, interfaces used to track jobs and tasks have colorization for status correlation, automatic refresh intervals, and more.

Security

Security is an important concern for enterprise software, and in the case of open source Hadoop, you need to be aware of some definite shortcomings when you put Hadoop into action. The good news is that BigInsights addresses these issues by securing access to the administrative interfaces and key Hadoop services.

The BigInsights administration console has been structured to act as a gateway to the cluster. It features enhanced security by supporting LDAP authentication. LDAP and reverse-proxy support help administrators restrict access to authorized users. In addition, clients outside the cluster must use REST HTTP access. In contrast, Apache Hadoop has open ports on every node in the cluster. The more ports you have to have open (and there are a lot of them in open source Hadoop), the less secure the environment, because the surface area isn't minimized.

BigInsights can be configured to communicate with a Lightweight Directory Access Protocol (LDAP) credentials server for authentication. All communication between the console and the LDAP server occurs using LDAP (by default) or both LDAP and LDAPS (LDAP over HTTPS). The BigInsights installer helps you to define mappings between your LDAP users and groups and the four BigInsights roles (System Administrator, Data Administrator, Application Administrator, and User). After BigInsights has been installed, you can add or remove users from the LDAP groups to grant or revoke access to various console functions.

Kerberos security is integrated in a competing Hadoop vendor that merely offers services and some operational tooling, but does not support alternative authentication protocols (other than Active Directory). BigInsights uses LDAP as the default authentication protocol. The development team has emphasized the use of LDAP because, as compared to Kerberos and other alternatives, it's a much simpler protocol to install and configure. That said, BigInsights does provide pluggable authentication support, enabling alternatives such as Kerberos.

BigInsights, with the use of GPFS-SNC, offers security that is less complex and inherently more secure than HDFS-based alternatives. Again, because GPFS-SNC is a kernel-level file system, it's naturally aware of users and groups defined in the operating system.

Upcoming changes to Apache Hadoop have improved security for HDFS, but because HDFS is not a kernel-level file system, this still will require additional complexity and processing overhead. As such, the experience IBM has with locking down the enterprise, which is baked into BigInsights, allows you to build a more secure, robust, and more easily maintained multitenant solution.

Enterprise Integration

A key component of IBM's vision for Big Data is the importance of integrating any relevant data sources; you're not suddenly going to have a Hadoop engine meet all your storage and processing needs. You have other investments in the enterprise, and being able leverage your assets (the whole left hand–right hand baseball analogy from Chapter 2 in this book) is going to be key. Enterprise integration is another area IBM understands very well. As such, BigInsights supports data exchange with a number of sources, including Netezza; DB2 for Linux, UNIX, and Windows; other relational data stores via a Java Database Connectivity (JDBC) interface; InfoSphere Streams; InfoSphere Information Server (specifically, Data Stage); R Statistical Analysis Applications; and more.

Netezza

BigInsights includes a connector that enables bidirectional data exchange between a BigInsights cluster and Netezza appliance. The Netezza Adapter is implemented as a Jaql module, which lets you leverage the simplicity and flexibility of Jaql in your database interactions.

The Netezza Adapter supports splitting tables (a concept similar to splitting files). This entails partitioning the table and assigning each divided portion to a specific mapper. This way, your SQL statements can be processed in parallel.

The Netezza Adapter leverages Netezza's external table feature, which you can think of as a materialized external UNIX pipe. External tables use JDBC. In this scenario, each mapper acts as a database client. Basically, a mapper (as a client) will connect to the Netezza database and start a read from a UNIX file that's created by the Netezza infrastructure.

DB2 for Linux, UNIX, and Windows

You can exchange data between BigInsights and DB2 for Linux, UNIX, and Windows in two ways: from your DB2 server through a set of BigInsights user defined functions (UDFs) or from your BigInsights cluster through the JDBC module (described in the next section).

The integration between BigInsights and DB2 has two main components: a set of DB2 UDFs and a Jaql server (to listen for requests from DB2) on the BigInsights cluster. The Jaql server is a middleware component that can accept Jaql query processing requests from a DB2 9.5 server or later. Specifically, the Jaql server can accept the following kinds of Jaql queries from a DB2 server:

- Read data from the BigInsights cluster.
- Upload (or remove) modules of Jaql code in the BigInsights cluster.
- Submit Jaql jobs (which can refer to modules you previously uploaded from DB2) to be run on the BigInsights cluster.

Running these BigInsights functions from a DB2 server gives you an easy way to integrate with Hadoop from your traditional application framework. With these functions, database applications (which are otherwise Hadoop-unaware) can access data in a BigInsights cluster using the same SQL interface they use to get relational data out of DB2. Such applications can now leverage the parallelism and scale of a BigInsights cluster without requiring extra configuration or other overhead. Although this approach incurs additional performance overhead as compared to a conventional Hadoop application, it is a very useful way to integrate Big Data processing into your existing IT application infrastructure.

JDBC Module

The Jaql JDBC module enables you to read and write data from any relational database that has a standard JDBC driver. This means you can easily exchange data and issue SQL statements with every major database warehouse product in the market today.

With Jaql's MapReduce integration, each map task can access a specific part of a table, enabling SQL statements to be processed in parallel for partitioned databases.

InfoSphere Streams

As you'll discover in Chapter 6, Streams is the IBM solution for real-time analytics on streaming data. Streams includes a sink adapter for BigInsights, which lets you store streaming data directly into your BigInsights cluster. Streams also includes a source adapter for BigInsights, which lets Streams applications read data from the cluster. The integration between BigInsights and Streams raises a number of interesting possibilities. At a high level, you would be able to create an infrastructure to respond to events in real time (as the data is being processed by Streams), while using a wealth of existing data (stored and analyzed by BigInsights) to inform the response. You could also use Streams as a large-scale data ingest engine to filter, decorate, or otherwise manipulate a stream of data to be stored in the BigInsights cluster.

Using the BigInsights sink adapter, a Streams application can write a control file to the BigInsights cluster. BigInsights can be configured to respond to the appearance of such a file so that it would trigger a deeper analytics operation to be run in the cluster. For more advanced scenarios, the trigger file from Streams could also contain query parameters to customize the analysis in BigInsights.

Streams and BigInsights share the same text analytics capabilities through the Advanced Text Analytics Toolkit (known initially by its IBM Research codename, SystemT). In addition, both products share a common end user web interface for parameterizing and running workloads. Future releases will feature additional alignment in analytic tooling.

InfoSphere DataStage

DataStage is an extract, transform, and load (ETL) platform that is capable of integrating high volumes of data across a wide variety of data sources and target applications. Expanding its role as a data integration agent, DataStage has been extended to work with BigInsights and can push and pull data to and from BigInsights clusters.

The DataStage connector to BigInsights integrates with both the HDFS and GPFS-SNC file systems, taking advantage of the clustered architecture so that any bulk writes to the same file are done in parallel. In the case of

GPFS-SNC, bulk writes can be done in parallel as well (because GPFS-SNC, unlike HDFS, is fully POSIX-compliant).

The result of DataStage integration is that BigInsights can now quickly exchange data with any other software product able to connect with DataStage. Plans are in place for even tighter connections between Information Server and BigInsights, such as the ability to choreograph BigInsights jobs from DataStage, making powerful and flexible ETL scenarios possible. In addition, designs are in place to extend Information Server information profiling and governance capabilities to include BigInsights.

R Statistical Analysis Applications

BigInsights includes an R module for Jaql, which enables you to integrate the R Project (see www.r-project.org for more information) for Statistical Computing into your Jaql queries. Your R queries can then benefit from Jaql's MapReduce capabilities and run R computations in parallel.

Improved Workload Scheduling: Intelligent Scheduler

Open source Hadoop ships with a rudimentary first in first out (FIFO) scheduler and a pluggable architecture supporting alternative scheduling options. Two pluggable scheduling tools are available through the Apache Hadoop project: the *Fair Scheduler* and the *Capacity Scheduler*. These schedulers are similar in that they enable a minimum level of resources to be available for smaller jobs to avoid starvation. (The Fair Scheduler is included in BigInsights while the Capacity Scheduler is not.) These schedulers do not provide adequate controls to ensure optimal cluster performance or offer administrators the flexibility they need to implement customizable workload management requirements. For example, FAIR is pretty good at ensuring resources are applied to workloads, but it doesn't give you SLA-like granular controls.

Performance experts in IBM Research have studied the workload scheduling problems in Hadoop and have crafted a solution called the *Intelligent Scheduler* (previously known as the *FLEX* scheduler). This scheduler extends the Fair Scheduler and manipulates it by constantly altering the minimum number of slots assigned to jobs. The Intelligent Scheduler includes a variety of metrics you can use to optimize your workloads. These metrics can be chosen by an administrator on a cluster-wide basis, or by individual users on a job-specific basis. You can optionally weight these metrics to balance competing priorities, minimize the sum of all the individual job metrics, or maximize the sum of all of them.

The following are examples of the Intelligent Scheduler controls you can use to tune your workloads:

`average response time`	The scheduler allocates maximum resources to small jobs, ensuring that these jobs are completed quickly.
`maximum stretch`	Jobs are allocated resources in proportion to the amount of resources they need. In other words, big jobs are higher in priority.
`user priority`	Jobs for a particular user are allocated the maximum amount of resources until complete.

Adaptive MapReduce

IBM Research workload management and performance experts have been working with Hadoop extensively, identifying opportunities for performance optimizations. IBM Research has developed a concept called *Adaptive MapReduce*, which extends Hadoop by making individual mappers self-aware and aware of other mappers. This approach enables individual map tasks to adapt to their environment and make efficient decisions.

When a MapReduce job is about to begin, Hadoop divides the data into many pieces, called *splits*. Each split is assigned a single mapper. To ensure a balanced workload, these mappers are deployed in waves, and new mappers start once old mappers finish processing their splits. In this model, a small split size means more mappers, which helps ensure balanced workloads and minimizes failure costs. However, smaller splits also result in increased cluster overhead due to the higher volumes of startup costs for each map task. For workloads with high startup costs for map tasks, larger split sizes tend to be more efficient. An adaptive approach to running map tasks gives BigInsights the best of both worlds.

One implementation of Adaptive MapReduce is the concept of an *adaptive mapper*. Adaptive Mappers extend the capabilities of conventional Hadoop mappers by tracking the state of file splits in a central repository. Each time an

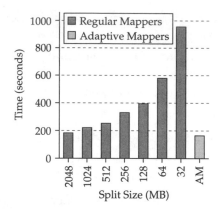

Figure 5-8 *Benchmarking a set-similarity join workload with high-map and task startup costs with Adaptive Mappers.*

Adaptive Mapper finishes processing a split, it consults this central repository and locks another split for processing until the job is completed. This means that for Adaptive Mappers, only a single wave of mappers is deployed, since the individual mappers remain open to consume additional splits. The performance cost of locking a new split is far less than the startup cost for a new mapper, which accounts for a significant increase in performance. Figure 5-8 shows the benchmark results for a set-similarity join workload, which had high map task startup costs that were mitigated by the use of Adaptive Mappers. The Adaptive Mappers result (see the AM bar) was based on a low split size of 32 MB. Only a single wave of mappers was used, so there were significant performance savings based on avoiding the startup costs for additional mappers.

For some workloads, any lack of balance could get magnified with larger split sizes, which would cause additional performance problems. When using Adaptive Mappers, you can, without penalty, avoid imbalanced workloads by tuning jobs to use a lower split size. Since there will only be a single wave of mappers, your workload will not be crippled by the mapper startup costs of many additional mappers. Figure 5-9 shows the benchmark results for a join query on TERASORT records, where an imbalance occurred between individual map tasks that led to an imbalanced workload for the higher split sizes. The Adaptive Mappers result (again, see the AM bar) was based on a low split size of 32 MB. Only a single wave of mappers was used, so there were significant performance savings based on the startup costs for additional mappers.

A number of additional Adaptive MapReduce performance optimization techniques are in development and will be released in future versions of BigInsights.

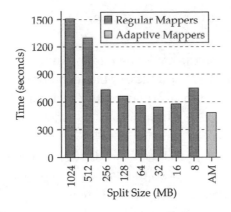

Figure 5-9 *Benchmark results for a join query on TERASORT records.*

Data Discovery and Visualization: BigSheets

Up until now in this chapter, we have been discussing foundational infrastructure aspects of BigInsights. Those are important features, which make Hadoop faster, more reliable, and more flexible for use in your enterprise. But the end goal of storing data is to get value out of it, which brings us to the BigInsights analytics capabilities. This is another major distinguishing feature of BigInsights, which makes it far more than just a Hadoop distribution—it is a platform for Big Data analytics. Unlike the core Apache Hadoop components or competitive bundled Hadoop distributions, BigInsights includes tooling for visualizing and performing analytics on large sets of varied data. Through all of its analytics capabilities, BigInsights hides the complexity of MapReduce, which enables your analysts to focus on analysis, not the intricacies of programming parallel applications.

Although Hadoop makes analyzing Big Data possible, you need to be a programmer with a good understanding of the MapReduce paradigm to explore the data. BigInsights includes a browser-based visualization tool called Big-Sheets, which enables line of business users to harness the power of Hadoop using a familiar spreadsheet interface. BigSheets requires no programming or special administration. If you can use a spreadsheet, you can use BigSheets to perform analysis on vast amounts of data, in any structure.

Three easy steps are involved in using BigSheets to perform Big Data analysis:

1. *Collect data.* You can collect data from multiple sources, including crawling the Web, local files, or files on your network. Multiple protocols and formats are supported, including HTTP, HDFS, Amazon S3 Native File System (s3n), and Amazon S3 Block File System (s3). When crawling the Web, you can specify the web pages you want to crawl and the crawl depth (for instance, a crawl depth of two gathers data from the starting web page and also the pages linked from the starting page). There is also a facility for extending BigSheets with custom plug-ins for importing data. For example, you could build a plug-in to harvest Twitter data and include it in your BigSheets collections.

2. *Extract and analyze data.* Once you have collected your information, you can see a sample of it in the spreadsheet interface, such as that shown in Figure 5-10. At this point, you can manipulate your data using the spreadsheet-type tools available in BigSheets. For example,

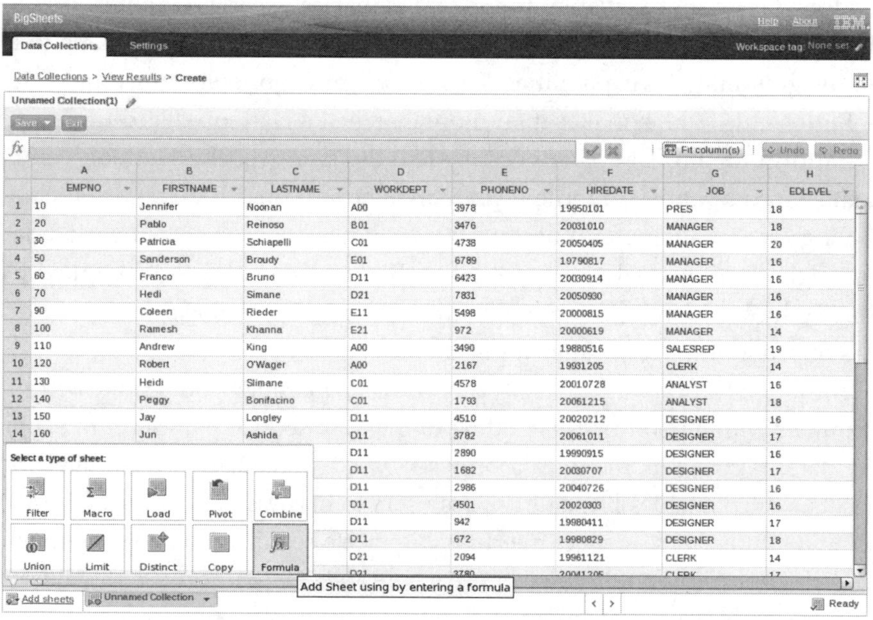

	A EMPNO	B FIRSTNAME	C LASTNAME	D WORKDEPT	E PHONENO	F HIREDATE	G JOB	H EDLEVEL
1	10	Jennifer	Noonan	A00	3978	19950101	PRES	18
2	20	Pablo	Reinoso	B01	3476	20031010	MANAGER	18
3	30	Patricia	Schiapelli	C01	4738	20050405	MANAGER	20
4	50	Sanderson	Broudy	E01	6789	19790817	MANAGER	16
5	60	Franco	Bruno	D11	6423	20030914	MANAGER	16
6	70	Hedi	Simane	D21	7831	20050930	MANAGER	16
7	90	Coleen	Rieder	E11	5498	20000815	MANAGER	16
8	100	Ramesh	Khanna	E21	972	20000619	MANAGER	14
9	110	Andrew	King	A00	3490	19880516	SALESREP	19
10	120	Robert	O'Wager	A00	2167	19931205	CLERK	14
11	130	Heidi	Slimane	C01	4578	20010728	ANALYST	16
12	140	Peggy	Bonifacino	C01	1793	20061215	ANALYST	18
13	150	Jay	Longley	D11	4510	20020212	DESIGNER	16
14	160	Jun	Ashida	D11	3782	20061011	DESIGNER	17
				D11	2890	19990915	DESIGNER	16
				D11	1682	20030707	DESIGNER	17
				D11	2986	20040726	DESIGNER	16
				D11	4501	20020303	DESIGNER	16
				D11	942	19980411	DESIGNER	17
				D11	672	19980829	DESIGNER	18
				D21	2094	19961121	CLERK	14

Select a type of sheet: Filter, Macro, Load, Pivot, Combine, Union, Limit, Distinct, Copy, Formula

Add Sheet using by entering a formula

Figure 5-10 *Analyze data in BigSheets*

you can combine columns from different collections, run formulas, or filter data. You can also include custom plug-ins for macros that you use against your data collections. While you build your sheets and refine your analysis, you can see the interim results in the sample data. It is only when you click the Run button that your analysis is applied to your complete data collection. Since your data could range from gigabytes to terabytes to petabytes, working iteratively with a small data set is the best approach.

3. *Explore and visualize data.* After running the analysis from your sheets against your data, you can apply visualizations to help you make sense of your data. BigSheets provides the following visualization tools:

- **Tag Cloud**   Shows word frequencies; the bigger the word, the more frequently it exists in the sheet. See Figure 5-11 for an example.

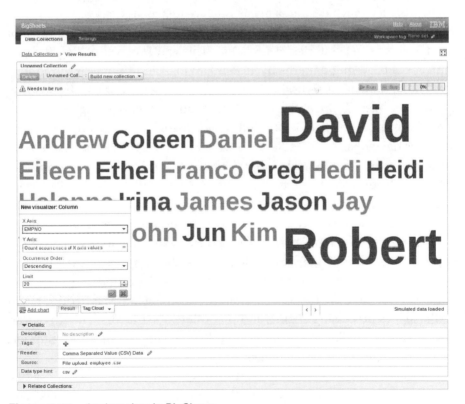

Figure 5-11   *Analyze data in BigSheets*

- **Pie Chart**   Shows proportional relationships, where the relative size of the slice represents its proportion of the data.

- **Map**   Shows data values overlaid onto either a map of the world or a map of the United States.

- **Heat Map**   Similar to the Map, but with the additional dimension of showing the relative intensity of the data values overlaid onto the Map.

- **Bar Chart**   Shows the frequency of values for a specified column.

BigSheets is fully extensible with its visualization tools. As such, you can include custom plug-ins for specialized renderings for your data.

Advanced Text Analytics Toolkit

While BigSheets is geared for the line-of-business user, BigInsights includes capabilities for much deeper analysis, such as text analytics.

Text analytics is growing in importance as businesses strive to gain insight from their vast repositories of text data. This can involve looking for customer web browsing patterns in clickstream log files, finding fraud indicators through email analytics, or assessing customer sentiment from social media messages. To meet these challenges, and more, BigInsights includes the Advanced Text Analytics Toolkit, which features a text analytics engine that was developed by IBM Research starting in 2004 under the codename SystemT. Since then, the Advanced Text Analytics Toolkit has been under continual development and its engine has been included in many IBM products, including Lotus Notes, IBM eDiscovery Analyzer, Cognos Consumer Insight, InfoSphere Warehouse, and more. Up to this point, the Advanced Text Analytics Toolkit has been released only as an embedded text analytics engine, hidden from end users. In BigInsights, the Advanced Text Analytics Toolkit is being made available as a text analytics platform that includes developer tools, an easy-to-use text analytics language, a MapReduce-ready text analytics processing engine, and prebuilt text extractors. The Advanced Text Analytics Toolkit also includes multilingual support, including support for double-byte character languages.

The goal of text analysis is to read unstructured text and distill insights. For example, a text analysis application can read a paragraph of text and derive structured information based on various rules. These rules are defined in *extractors*, which can, for instance, identify a person's name within a text field. Consider the following text:

```
In the 2010 World Cup of Soccer, the team from the
Netherlands distinguished themselves well, losing to
Spain 1-0 in the Final. Early in the second half, Dutch
striker Arjen Robben almost changed the tide of the
game on a breakaway, only to have the ball deflected
by Spanish keeper, Iker Casillas. Near the end of
regulation time, winger Andres Iniesta scored, winning
Spain the World Cup.
```

The product of these extractors is a set of annotated text, as shown in the underlined text in this passage.

Following is the structured data derived from this example text:

Name	Position	Country
Arjen Robben	Striker	Netherlands
Iker Casillas	Goalkeeper	Spain
Andres Iniesta	Winger	Spain

In the development of extractors and applications where the extractors work together, the challenge is to ensure the accuracy of the results. Accuracy can be broken down into two factors: *precision*, which is the percentage of items in the result set that are relevant (are the results you're getting valid?), and *recall*, which is the percentage of relevant results that are retrieved from the text (are all the valid strings from the original text showing up?). As analysts develop their extractors and applications, they iteratively make refinements to fine-tune their precision and recall rates.

Current alternative approaches and infrastructure for text analytics present challenges for analysts, as they tend to perform poorly (in terms of both accuracy and speed) and they are difficult to use. These alternative approaches rely on the raw text flowing only forward through a system of extractors and filters. This is an inflexible and inefficient approach, often resulting in redundant processing. This is because extractors applied later in the workflow might have done some processing already completed carlier. Existing toolkits are also limited in their expressiveness (specifically, the degree of granularity that's possible with their queries), which results in analysts having to develop custom code. This, in turn, leads to more delays, complexity, and difficulty in refining the accuracy of your result set (precision and recall).

The BigInsights Advanced Text Analytics Toolkit offers a robust and flexible approach for text analytics. The core of the Advanced Text Analytics Toolkit is its Annotator Query Language (AQL), a fully declarative text analytics language, which means there are no "black boxes" or modules that can't be customized. In other words, everything is coded using the same semantics and is subject to the same optimization rules. This results in a text analytics language that is both highly expressive and very fast. To the best of our knowledge, there are no other fully declarative text analytics languages available on the market today. You'll find high-level and medium-level declarative languages, but they all make use of locked-up black-box modules that cannot be customized, which restricts flexibility and are difficult to optimize for performance.

AQL provides an SQL-like language for building extractors. It's highly expressive and flexible, while providing familiar syntax. For example, the following AQL code defines rules to extract a person's name and telephone number.

```
create view PersonPhone as select P.name as person,
N.number as phone
from Person P, Phone PN, Sentence S where Follows(P.
name. PN.number, 0, 30)
    and Contains(S.sentence, P.name) and Contains(S.
sentence, PN.number)
    and ContainsRegex(/\b(phone|at)\b/, SpanBetween(P.
name, PN.number));
```

Figure 5-12 shows a visual representation of the extractor defined in the previous code block.

The Advanced Text Analytics Toolkit includes Eclipse plug-ins to enhance analyst productivity. When writing AQL code, the editor features syntax highlighting and automatic detection of syntax errors (see Figure 5-13).

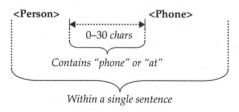

Figure 5-12 *Visual expression of extractor rules*

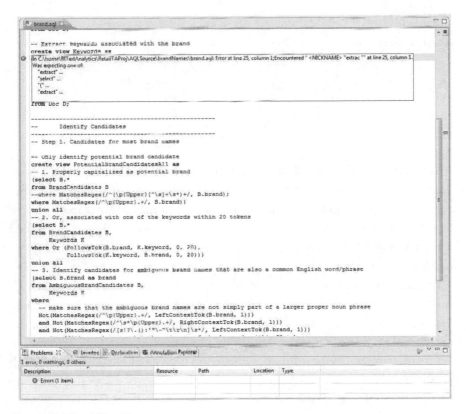

Figure 5-13 *AQL editor*

Also included is a facility to test extractors against a subset of data. This is important for analysts as they refine the precision and recall of their extractors. Testing their logic against the complete data sets, which could range up to petabytes of volume, would be highly inefficient and wasteful.

A major challenge for analysts is determining the lineage of changes that have been applied to text. It can be difficult to discern which extractors need to be adjusted to tweak the resulting annotations. To aid in this, the Provenance viewer, shown in Figure 5-14, features an interactive visualization, displaying exactly which rules influence the resulting annotations.

An additional productivity tool to aid analysts to get up and running quickly is the inclusion of a prebuilt extractor library with the Advanced Text Analytics Toolkit. Included are extractors for the following:

Acquisition	Address	Alliance
AnalystEarningsEstimate	City	CompanyEarningsAnnouncement
CompanyEarningsGuidance	Continent Country	ZipCode
County	DateTime	EmailAddress
JointVenture	Location	Merger
NotesEmailAddress	Organization	Person
PhoneNumber	StateOrProvince	URL

The fully declarative nature of AQL enables its code to be highly optimized. In contrast with the more rigid approaches to text analytics frameworks described earlier, the AQL optimizer determines order of execution of the extractor instructions for maximum efficiency. As a result, the Advanced Text Analytics Toolkit has delivered benchmark results up to ten times faster than leading alternative frameworks (see Figure 5-15).

When coupled with the speed and enterprise stability of BigInsights, the Advanced Text Analytics Toolkit represents an unparalleled value proposition. The details of the integration with BigInsights (described in Figure 5-16) are transparent to the text analytics developer. Once the finished AQL is compiled and then optimized for performance, the result is an Analytics Operator Graph (AOG) file. This AOG can be submitted to BigInsights as an analytics job through the BigInsights web console. Once submitted, this AOG is distributed

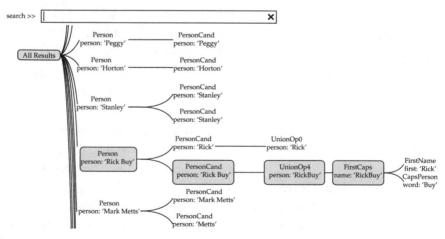

Figure 5-14 *Provenance viewer*

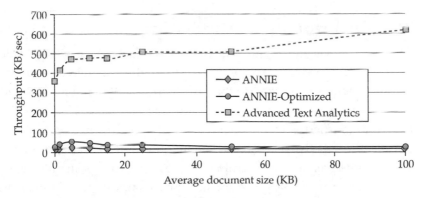

Figure 5-15 *Advanced Text Analytics Toolkit performance benchmark*

with every mapper to be executed on the BigInsights cluster. Once the job starts, each mapper then executes Jaql code to instantiate its own Advanced Text Analytics Toolkit runtime and applies the AOG file. The text from each mapper's file split is run through the toolkit's runtime, and an annotated document stream is passed back as a result set.

When you add up all its capabilities, the BigInsights Advanced Text Analytics Toolkit gives you everything you need to develop text analytics applications to help you get value out of extreme volumes of text data. Not only is there extensive tooling to support large-scale text analytics development, but the resulting code is highly optimized and easily deployable on a Hadoop cluster.

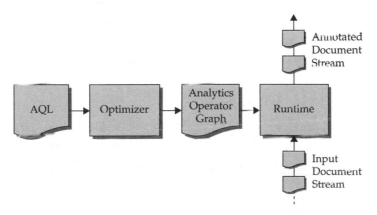

Figure 5-16 *Integration of Advanced Text Analytics Toolkit with BigInsights*

Machine Learning Analytics

In 2012, we believe BigInsights will include a Machine Learning Toolkit, which was developed by IBM Research under the codename SystemML. (Disclaimer: there is no guarantee that this feature will debut in 2012, but if we had to bet, we'd say you will see it sooner than later.) This provides a platform for statisticians and mathematicians to conduct high-performance statistical and predictive analysis on data in a BigInsights Hadoop cluster. It includes a high-level machine learning language, which is semantically similar to R (the open source language for statistical computing) and can be used by analysts to apply statistical models to their data processing. A wealth of precanned data mining algorithms and statistical models are included as well and are ready for customization.

The Machine Learning Toolkit includes an engine that converts the statistical workloads expressed in machine learning language into parallelized MapReduce code, so it hides this complexity from analysts. In short, analysts don't need to be Java programmers, and they don't need to factor MapReduce into their analytics applications.

The Machine Learning Toolkit was developed in IBM Research by a team of performance experts, PhD statisticians, and PhD mathematicians. Their primary goals were high performance and ease of use for analysts needing to perform complex statistical analysis in a Hadoop context. As such, this toolkit features optimization techniques for the generation of low-level MapReduce execution plans. This enables statistical jobs to feature orders of magnitude performance improvements, as compared to algorithms directly implemented in MapReduce. Not only do analysts not need to apply MapReduce coding techniques to their analytics applications, but the machine learning code they write is highly optimized for excellent Hadoop performance.

Large-Scale Indexing

To support its analytics toolkits, BigInsights includes a framework for building large-scale indexing and search solutions, called *BigIndex*. The indexing component includes modules for indexing over Hadoop, as well as optimizing, merging, and replicating indexes. The search component includes modules for programmable search, faceted search, and searching an index in local and distributed deployments. Especially in the case of text analytics, a robust index is vital to ensure good performance of analytics workloads.

BigIndex is built on top of the open source Apache Lucene search library and the IBM Lucene Extension Library (ILEL). IBM is a leading contributor to the Lucene project and has committed a number of Lucene enhancements through ILEL. Because of its versatile nature as an indexing engine, the technologies used in BigIndex are deployed in a number of products. In addition to BigInsights, it's included in Lotus Connections, IBM Content Analyzer, and Cognos Consumer Insight, to name a few. IBM also uses BigIndex to drive its Intranet search engine. (This project, known as Gumshoe, is documented heavily in the book *Hadoop in Action*, by Chuck Lam [Manning Publications, 2010].)

The goals for BigIndex are to provide large-scale indexing and search capabilities that leverage and integrate with BigInsights. For Big Data analytics applications, this means being able to search through hundreds of terabytes of data, while maintaining subsecond search response times. One key way BigIndex accomplishes this is by using various targeted search distribution architectures to support the different kinds of search activities demanded in a Big Data context. BigIndex can build the following types of indexes:

- **Partitioned index**   This kind of index is partitioned into separate indices by a metadata field (for example, a customer ID or date). A search is usually performed only on one of these indices, so the query can be routed to the appropriate index by a runtime query dispatcher.

- **Distributed index**   The index is distributed into shards, where the collection of shards together represents one logical index. Each search is evaluated against all shards, which effectively parallelizes the index key lookups.

- **Real-time index**   Data from real-time sources (for example, Twitter) is added to an index in near real time. The data is analyzed in parallel, and the index is updated when analysis is complete.

Figure 5-17 depicts a deployment of BigIndex, where indexing is done using a BigInsights cluster and search is done in its own shard cluster.

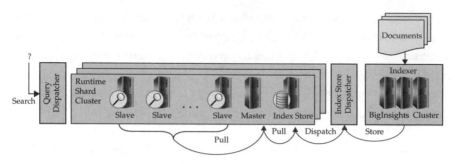

Figure 5-17   *Distributed BigIndex deployment*

The following steps are involved in generating and deploying an index for the kind of distributed environment shown in Figure 5-17:

1. **Data ingest**   Documents are ingested into the BigInsights cluster. This can be done through any available means—for instance, a flow of log files ingested through Flume, diagnostic data processed by Streams, or a Twitter feed stored in HDFS or GPFS-SNC.

2. **Data parsing**   Parse the documents to select fields that need to be indexed. It is important for the parsing algorithms to be selective: there needs to be a balance between good coverage (indexing fields on which users will search) and quantity (as more fields are indexed, performance slows). Text analytics can be used here, if needed.

3. **Data faceting**   Identify how the current documents relate to others by isolating and extracting the facets (such as categories) that users might want to use to narrow and drill down into their search results—for example, year, month, date; or country, state, city.

4. **Data indexing**   This indexing is based on a Lucene text index, but with many extensions. The documents are indexed using Hadoop by an *Indexer*, which is deployed as a MapReduce job. Two kinds of indexes are generated: a single Lucene index and a distributed index (which is composed of multiple Lucene indexes representing individual indices). Faceted indexing capability is integrated with both the single Lucene index and the distributed index.

5. **Index merging**   Once generated, the index is dispatched to the Runtime Shard Cluster for storage. The Master pulls the indexes

from the Index Store and merges them with its local index. This is not like a regular database index, where you can insert or delete values as needed. This index is an optimized data structure. As a result, incremental changes need to be merged into this structure.

6. **Index replication**   The slave processors replicate the index updates from the Master and are ready to serve search queries from users.

7. **Index searching**   BigIndex exposes its distributed search functionality through multiple interfaces, including a Java API, a scripting language using Jaql, and REST-like HTTP APIs.

BigInsights Summed Up

As the sum of the many parts described in this chapter, BigInsights represents a fast, robust, and easy-to-use platform for analytics on Big Data at rest. With our graphical installation, configuration, and administrative tools, management of the cluster is easy. By storing your data using GPFS-SNC, you gain performance improvements, but also high availability and flexibility in maintaining your data. The inclusion of the IBM LZO compression module enables you to compress your data with a high-performance algorithm, without licensing hassles. There are additional performance features, such as Adaptive MapReduce, and the Intelligent Scheduler, which helps you maintain reliable service level agreements with a user base that will come to depend on your Big Data analytics. And speaking of analytics, BigInsights provides capabilities for a wide range of users. For line-of-business users, BigSheets is a simple tool geared to create visualizations on large volumes of data. And for deeper analytics, BigInsights provides an industry-leading text analytics tookit and engine. And in the near future, the IBM Research Machine Learning Analytics Toolkit will be available as well. We think this represents an incredible story, which is unparalleled in the IT industry. Through BigInsights, you get a complete analytics solution, supported by the world's largest corporate research organization, a deep development team, and IBM's global support network.

6

IBM InfoSphere Streams: Analytics for Big Data in Motion

Now that you've read about how IBM uniquely handles the largest data analytical problems in a Hadoop environment that's hardened for the enterprise, let's turn our attention to the other side of the IBM Big Data story: analytics for data in motion. Using BigInsights gives you a competitive advantage by helping you with the ocean of information out there, and IBM InfoSphere Streams (Streams) gives you insights from the Niagara Falls of data flowing through your environment. You can either tap into that flow to gain time-sensitive competitive advantages for your business, or you can be like most people at Niagara Falls, and simply watch in awe as the mighty river flows past. This is where Streams comes in. Its design lets you leverage massively parallel processing (MPP) techniques to analyze data while it is streaming, *so you can* understand what is happening in real time and take action, make better decisions, and improve outcomes.

Before we delve into this chapter, let's start by clarifying what we mean by *Streams* and *streams*; the capitalized version refers to the IBM InfoSphere Streams product, and the lowercase version refers to a stream of data. With that in mind, let's look at the basics of Streams, its use cases, and some of the technical underpinnings that define how it works.

InfoSphere Streams Basics

Streams is a powerful analytic computing platform that delivers a platform for analyzing data in real time with micro-latency. Rather than gathering large quantities of data, manipulating the data, storing it on disk, and then analyzing it, as would be the case with BigInsights (in other words, analytics on data at rest), Streams allows you to apply the analytics on the data in motion. In Streams, data flows through operators that have the ability to manipulate the data stream (which can comprise millions of events per second), and in-flight analysis is performed on the data. This analysis can trigger events to enable businesses to leverage just-in-time intelligence to perform actions in real time ultimately yielding better results for the business. After flowing the data through the analytics, Streams provides operators to store the data into various locations (including BigInsights or a data warehouse among others) or just toss out the data if it is deemed to be of no value by the in-flight analysis (either because it wasn't interesting data or the the data has served its purpose and doesn't have persistence requirements).

If you are already familiar with Complex Event Processing (CEP) systems, you might see some similarities in Streams. However, Streams is designed to be much more scalable and is able to support a much higher data flow rate than other systems. In addition, you will see how Streams has much higher enterprise-level characteristics, including high availability, a rich application development toolset, and advanced scheduling.

You can think of a stream as a series of connected operators. The initial set of operators (or a single operator) are typically referred to as *source operators*. These operators read the input stream and in turn send the data downstream. The intermediate steps comprise various operators that perform specific actions. Finally, for every way into the in-motion analytics platform, there are multiple ways out, and in Streams, these outputs are called *sink operators* (like the water that flows out of the tap and into your kitchen sink). We'll describe all of these operators in detail later in this chapter.

We refer to Streams as a platform because you can build or customize Streams in almost any possible way to deliver applications that solve business problems; of course, it's an enterprise capable platform because each of these operators can be run on a separate server in your cluster to improve availability, scalability, and performance. For example, Streams provides a rich tooling environment to help you design your streaming applications (covered later in this chapter). Another nice thing is that Streams shares the same Text Analytics Toolkit with BigInsights, allowing you to reuse skills and code snippets across your entire Big Data platform. When you're ready to deploy your streaming application, Streams autonomically decides, at runtime, where to run the processing elements (PEs) based on cluster-based load balancing and availability metrics, allowing it to reconfigure operators to run on other servers to ensure the continuity of the stream in the event of server or software failures. You can also programmatically specify which operators run on which servers and run your streams logic on specific servers.

This autonomic streaming and customizable platform allows you to increase the number of servers performing analysis on the stream simply by adding additional servers and assigning operators to run on those servers. The Streams infrastructure ensures that the data flows successfully from one operator to another, whether the operators are running on distinct servers or on the same server: This provides a high degree of agility and flexibility to start small and grow the platform as needed.

Much like BigInsights, Streams is ideally suited not only for structured data, but for the other 80 percent of the data as well—the nontraditional semistructured or unstructured data coming from sensors, voice, text, video, financial, and many other high-volume sources. Finally, since Streams and BigInsights are part of the IBM Big Data platform, you'll find enormous efficiencies in which the analytics you build for in-motion or at-rest Big Data can be shared. For example, the extractors built from the Text Analytic Toolkit can be deployed in Streams or BigInsights.

Industry Use Cases for InfoSphere Streams

To give you some insight into how Streams technology can fit into your environment, we thought we would provide some industry use case examples. Obviously, we can't cover every industry in such a short book, but we think this section will get you thinking about the breadth of possibilities that Streams technology can offer your environment (get ready because your brain is going to shift into overdrive with excitement).

Financial Services Sector (FSS)

The financial services sector and its suboperations are a prime example for which the analysis of streaming data can provide a competitive advantage (as well as regulatory oversight, depending on your business). The ability to analyze high volumes of trading and market data, at ultra low latencies, across multiple markets and countries simultaneously, can offer companies the microsecond reaction times that can make the difference between profit and loss via arbitrage trading and book of business risk analysis (for example, how does such a transaction occurring at this very moment add to the firm's risk position?).

Streams can also be used by FSS companies for real-time trade monitoring and fraud detection. For example, Algo Trading supports average throughput rates of about 12.7 million option market messages per second and generates trade recommendations for its customers with a latency of 130 microseconds. As discussed later in this chapter, there are even adapters integrated into Streams that provide direct connectivity via the ubiquitous Financial Information eXchange (FIX) gateways with a function-rich library to help calculate theoretical Put and Call option values. Streams can even leverage multiple types of inputs. For example, you could use Streams to analyze impeding weather patterns and their impact on security prices as part of a short-term hold position decision.

Similarly, real-time fraud detection can also be used by credit card companies and retailers to deliver fraud and multi-party fraud detection (as well as to identify real-time up-sell or cross-sell opportunities).

Health and Life Sciences

Healthcare equipment is designed to produce diagnostic data at a rapid rate. From electrocardiograms, to temperature and blood pressure measuring devices, to blood oxygen sensors, and much more, medical diagnostic equipment produces a vast array of data. Harnessing this data and analyzing it in real time delivers benefits unlike any other industry; that is, in addition to providing a company with a competitive advantage, Streams usage in healthcare is helping to save lives.

For example, the University of Ontario Institute of Technology (UOIT) is building a smarter hospital in Toronto and leveraging Streams to deliver a neonatal critical care unit that monitors the health of what we'll lovingly and affectionately call these little miracles, "data babies." These babies continually generate data in a neonatal ward: every heartbeat, every breath, every anomaly, and more. With more than 1000 pieces of unique information per second of medical diagnostic information, the Streams platform is used as an early warning system that helps doctors find new ways to avoid life-threatening infections up to 24 hours sooner than in the past. There is a synergistic effect at play here, too. It could be the case that separately monitored streams absolutely fall within normal parameters (blood pressure, heart rate, and so on); however, the combination of several streams with some specific value ranges can turn out to be the predictor of impending illness. Because Streams is performing analytics on moving data instead of just looking for out of bound values, it not only has the potential to save lives, but it also helps drive down the cost of healthcare (check it out at: http://www.youtube.com/watch?v=QVbnrlqWG5I).

Telecommunications

The quantity of call detail records (CDRs) that telecommunications (telco) companies have to manage is staggering. This information is not only useful for providing accurate customer billing, but a wealth of information can be gleaned from CDR analysis performed in near real time. For example, CDR analysis can help to prevent customer loss by analyzing the access patterns of "group leaders" in their social networks. These group leaders are people who might be in a position to affect the tendencies of their contacts to move from one service provider to another. Through a combination of traditional and social media analysis, Streams can help you identify these individuals, the networks to which they belong, and on whom they have influence.

Streams can also be used to power up a real-time analytics processing (RTAP) campaign management solution to help boost campaign effectiveness, deliver a shorter time-to-market for new promotions and soft bundles, help find new revenue streams, and enrich churn analysis. For example, Globe Telecom leverages information gathered from its handsets to identify the optimal service promotion for each customer and the best time to deliver it, which has had profound effects on its business. Globe Telecom reduced from 10 months to 40 days the time-to-market for new services, increased sales significantly through real-time promotional engines, and more.

What's good for CDRs can also be applied to Internet Protocol Detail Records (IPDRs). IPDRs provide information about Internet Protocol (IP)–based service usage and other activities that can be used by operational support to determine the quality of the network and detect issues that might require maintenance before they lead to a breakdown in network equipment. (Of course, this same use case can be applied to CDRs.) Just how real-time and low-latency is Streams when it comes to CDR and IPDR processing? We've seen supported peak throughput rates of some detail records equal to 500,000 per second, with more than 6 billion detail records analyzed per day (yes, you read that rate right) on more than 4 PBs (4000 TBs) of data per year; CDR processing with Streams technology has sustained rates of 1 GBps, and X-ray Diffraction (XRD) rates at 100 MBps. Truly, Streams is game changing technology.

Enforcement, Defense, Surveillance, and Cyber Security

Streams provides a huge opportunity for improved law enforcement and increased security, and offers unlimited potential when it comes to the kinds of applications that can be built in this space, such as real-time name recognition, identity analytics, situational awareness applications, multimodal surveillance, cyber security detection, wire taps, video surveillance, and face recognition. Corporations can also leverage streaming analytics to detect and prevent cyber attacks by streaming network and other system logs to stop intrusions or detect malicious activity anywhere in their networks.

TerraEchos uses InfoSphere Streams to provide covert sensor surveillance systems to enable companies with sensitive facilities to detect intruders before they even get near the buildings or other sensitive installations. They've been a recipient of a number of awards for their technology (the Frost and Sullivan Award for Innovative Product of the Year for their Fiber Optic Sensor System Boarder Application, among others). The latest version of Streams includes a brand new development framework, called Streams Processing Language (SPL), which allows them to deliver these kinds of applications 45 percent faster than ever before, making their capability, and the time it takes to deliver it, that much faster.

And the Rest We Don't Have Space for in This Book...

As we said, we can't possibly cover all the use cases and industries that a potent product such as Streams can help solve, so we'll cram in a couple more, with fewer details, here in this section.

Government agencies can leverage the broad real-time analytics capabilities of Streams to manage such things as wildfire risks through surveillance and weather prediction, as well as manage water quality and water consumption through real-time flow analysis. Several governments are also improving traffic flow in some of their most congested cities by leveraging GPS data transmitted via taxis, traffic flow cameras, and traffic sensors embedded in roadways to provide intelligent traffic management. This real-time analysis can help them predict traffic patterns and adjust traffic light timings to improve the flow of traffic, thereby increasing the productivity of their citizens by allowing them to get to and from work more efficiently.

The amount of data being generated in the utilities industry is growing at an explosive rate. Smart meters as well as sensors throughout modern energy grids are sending real-time information back to the utility companies at a staggering rate. The massive parallelism built into Streams allows this data to be analyzed in real time such that energy distributors and generators are able to modify the capacity of their electrical grids based on the changing demands of consumers. In addition, companies can include data on natural systems (such as weather or water management data) into the analytics stream to enable energy traders to meet client demand while at the same time predicting consumption (or lack of consumption) requirements to deliver competitive advantages and maximize company profits.

Manufacturers want more responsive, accurate, and data rich quality records and quality process controls (for example, in the microchip fabrication domain, but applicable to any industry) to better predict, avoid, and determine defined out of tolerance events and more. E-science domains such as space weather prediction, detection of transient events, and Synchrotron atomic research are other opportunities for Streams. From smarter grids, to text analysis, to "Who's talking to Whom?" analysis, and more, Streams use cases, as we said earlier, are nearly limitless.

How InfoSphere Streams Works

As mentioned, Streams is all about analytics on data in motion. You can think of a stream as somewhat like a set of dominoes in a line. When you push the first one over, you end up with a chain reaction (assuming you have lined everything up right) where the momentum of one falling domino is enough to start the next one falling and so on. If you are good, you can even have the line of dominoes split into several lines of simultaneously falling tiles and then merge them back together at some point down the line. In this way you have many dominoes falling in parallel, all feeding the momentum to the next dominoes in their line. (In case you are wondering, according to the Guinness Book of World Records, the greatest number of dominoes toppled by a group at one time is over 4.3 million.) Streams is similar in nature in that some data elements start off a flow which moves from operator to operator, with the output of one operator becoming the input for the next. Similarly, a record, or tuple, of data can be split into multiple streams and potentially joined back together downstream. The big difference of course is that with the game Dominoes, once a tile falls down, that's the end of it, whereas with Streams, the data continuously flows through the system at very high rates of speed, allowing you to analyze a never-ending flow of information continuously.

What's a Stream?

In a more technical sense, a stream is a *graph of nodes connected by edges.* Each node in the graph is an *operator* or *adapter* that will process the data within the stream in some way. Nodes can have zero or more inputs and zero or more outputs. The output (or outputs) from one node is connected to the input (or inputs) of another node or nodes. The edges of the graph that join the nodes together represent the stream of data moving between the operators. Figure 6-1 represents a simple stream graph that reads data from a file, sends the data to an operator known as a *functor* (this operator transforms incoming data in some programmatic manner), and then feeds that data to another operator. In this figure, the streamed data is fed to a *split operator*, which then feeds data to either a file sink or a database (depending on what goes on inside the split operator).

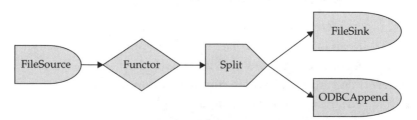

Figure 6-1   *A simple data stream that applies a transformation to some data and splits it into two possible outputs based on some predefined logic.*

Data flows through a stream in what are known as tuples. In a relational database sense, you can think of them as rows of data. However, when Streams works on semistructured and unstructured data, a tuple is an abstraction to represent a package of data. Think of a tuple as a set of attributes for a given object. Each element in the tuple contains the value for that attribute, which can be a character string, a number, a date, or even some sort of binary object.

Some operators work on an individual tuple, transforming the data and then passing it along. Other operators need to work on groups of tuples before they send out results. Consider, for example, a sort operation: you can't sort data by simply working on one tuple at a time. You must have a set of data that you can put into sorted order and then pass along. For this reason, some operators work on a *window* of data, which is essentially a set of tuples that are grouped together. The operator itself will define how many tuples are in the window based on the window expression inside the operator. For example, the operator can define a window to be the next N tuples that come into the operator, or it could define the window to be any tuple that enters the operator in the next M seconds. There are many other ways to define a window, and in fact some windows can be moving (so they define a sliding window of tuples) and others are more batching (they group together a set of tuples, empty the operator of all those tuples at some certain point in time or event, and subsequently group together the next set). We discuss windowing later in this chapter where we talk about the various operators, but it's an important concept to understand in that Streams *is not* just about manipulating one tuple at a time, but rather analyzing large sets of data in real time.

The Streams Processing Language

The Streams Processing Language (SPL) is a structured application development language that is used by Streams to create your applications. It's a higher generation and more productive programming framework for Streams than what was available in past releases; in fact, one customer claims up to 45 percent productivity improvements in their Streams applications because of the rich SPL. After all, technology is great, but if you can't quickly apply it to the business need at hand, what's the point?

Streams-based applications written in SPL are compiled using the Streams compiler, which turns them into binary (bin) executable code, which then runs in the Streams environment to accomplish the tasks on the various servers in the cluster. An SPL program is a text-based representation of the graph we discussed in the preceding section: It defines the sources, sinks, and operators between them that in turn define the stream processing and how each operator will behave within the stream. Later in this chapter, we talk about the application development tooling that makes building Streams applications simple. But in the same way children learn to multiply in their head before being allowed to use a calculator, we are going to look at SPL before showing you the easy application development tooling. For example, the following SPL code snippet represents a simple stream from one source, through a single operator, and to an eventual single sink:

```
composite toUpper {
   graph
     stream<rstring line> LineStream = FileSource() {
      param  file              : "input_file";
             format            : line;
     }
     stream<LineStream> upperedTxt = Functor(LineStream)
{
      output upperedTxt        : line = upper(line);
     }
     () as Sink = FileSink(upperedTxt) {
     param  file      : "/dev/stdout";
     format           : line;
     }
}
```

In this SPL snippet, the `FileSource` operator reads data from the specified file and puts it into a stream called `LineStream`. The operator in this case is called a `Functor` operator, which converts data from the stream to the uppercase text of that stream and puts that tuple on an output stream called `upperedTxt`. The `Sink` operator then reads the `upperedTxt` stream of data and sends it, in this case, to standard output.

This snippet represents the simplest stream with a single source, a single operation, and a single sink. Of course, the power of Streams is that it can run massively parallel jobs across large clusters of servers where each operator, or a group of operators, can be running on a separate server. But before we get into the enterprise class capabilities of Streams, let's look at the various adapters that are available with this product.

Source and Sink Adapters

It goes without saying that in order to perform analysis on a stream of data, the data has to enter the stream. Of course, a stream of data has to go somewhere when the analysis is done (even if somewhere is defined as a void where bits just get dumped into "nowhere"). Let's look at the most basic source adapters available to ingest data along with the most basic sink adapters to which data can be sent.

FileSource and FileSink

As the names imply, `FileSource` and `FileSink` are standard adapters used to read or write to a file. You use parameters to specify the name and location of the file used for the read or write operation. Another parameter identifies the format of the file's contents, which could be any of the following:

`txt`   Simple text files, where each tuple is a row in the file

`csv`   Files that contain comma-separated values

`bin`   Files that contain binary data tuples

`line`   Files that contain lines of text data

`block` An input stream made up of binary data blocks (much like a BLOB)

There are a number of other optional parameters that can be used to specify, for example, column separators, end of line markers, delimiters, and more.

TCPSource/UDPSource and TCPSink/UDPSink

The `TCPSource` and `TCPSink` adapters are the basic TCP adapters used in Streams to read and write to a socket. When you use these adapters, you specify the IP address (using either IPv4 or IPv6) along with the port, and the adapter will read from the socket and generate tuples into the stream. These adapters' parameters are the same as the `FileSource` and `FileSink` adapters in terms of the format of the data flow (`txt`, `csv`, and so on). The `UDPSource` and `UDPSink` adapters read from and write to a UDP socket in the same manner as the TCP-based adapters.

Export and Import

The `export` and `import` adapters work together within a stream. You can export data using the `export` adapter and assign the exported stream a `streamID`. Once the stream is assigned this ID, any other stream application in the same instance can import this data using the assigned `streamID`. Using `export` and `import` is a great way to stream data between applications running under the same Streams instance.

MetricsSink

The `MetricsSink` adapter is a very interesting and useful sink adapter because it allows you to set up a *named meter*, which is incremented whenever a tuple arrives at the sink. You can think of these meters as a gauge that you can monitor using Streams Studio or other tools. If you've ever driven over one of those traffic counters (those black rubber tubes that seem to have no purpose, rhyme, or reason for lying across an intersection or road) you've got the right idea, and while a traffic counter measures the flow of traffic through a point of interest, a `MetricsSink` can be used to monitor the volume and velocity of data flowing out of your data stream.

Operators

Quite simply, operators are at the heart of the Streams analytical engine. They take data from upstream adapters or other operators, manipulate that data, and then move the resulting tuples downstream to the next operator. In this section we discuss some of the more common Streams operators that can be strung together to build a Streams application.

Filter

The `filter` operator is similar to a filter in an actual water stream, or in your furnace or car: its purpose is to allow only some of the streaming contents to pass. A Streams `filter` operator removes tuples from a data stream based on a user-defined condition specified as a parameter to the operator. Once you've programmatically specified a condition, the first output port defined in the operator will receive any tuples that satisfy that condition. You can optionally specify a second output port, to receive any tuples that did not satisfy the specified condition. (If you're familiar with extract, transform, and load [ETL] flows, this is similar to a `match` and `discard` operation.)

Functor

The `functor` operator reads from the input stream, transforms them in some way, and sends those tuples to an output stream. The transformation applied can manipulate any of the elements in the stream. For example, you could extract a data element out of a stream and output the running total of that element for every tuple that comes through a specific functor operator.

Punctor

The punctor operator adds punctuation into the stream, which can then be used downstream to separate the stream into multiple windows. For example, suppose a stream reads a contact directory listing and processes the data flowing through that stream. You can keep a running count of last names in the contact directory by using the punctor operator to add a punctuation mark into the stream any time your application observes a change in the last name in the stream. You could then use this punctuation mark downstream in an aggregation functor operator to send out the running total for that name, later resetting the count back to 0 to start counting the next set of last names.

Sort

The sort operator is fairly easy to understand in that it simply outputs the same tuples it receives, but in a specified sorted order. This is the first operator we've discussed that uses a stream window specification. Think about it for a moment: if a stream represents a constant flow of data, how can you sort the data, because you don't know if the next tuple to arrive will need to be sorted to the first tuple you must send as output? To overcome this issue, Streams allows you to specify a window on which you want to operate. You can specify a window of tuples in a number of ways:

count   The number of tuples to include in the window

delta   Wait until a given attribute of an element in the stream has changed by a specified delta amount

time   The amount of time in seconds you want to wait to allow the window to fill up

punctuation The punctuation used to delimit the window, as defined by the punctor operator

In addition to specifying the windowing option, you must also specify the expression that defines how you want the data sorted (for example, sort by a given attribute in the stream). Once the window fills up, the sort operator will sort the tuples based on the element you specified, and sends those tuples to the output port in the defined sorted order (then it goes back to filling up the window again). By default, Streams sorts in ascending order, but you can also specify that you want a descending sort.

Join

As you've likely guessed, the `join` operator takes two streams, matches the tuples on a user-specified condition, and then sends the matches to an output stream. When a row arrives on one input stream, the matching attribute is compared to the tuples that already exist in the operating window of the second input stream to try to find a match. Just as in a relational database, several types of joins can be used, including `inner joins` (in which only matches will be passed on) and `outer joins` (which can pass on one of the stream tuples, even without a match in addition to matching tuples from both streams). As with the `sort` operator, you must specify a window of tuples to store in each stream in order to join them.

Aggregate

The `aggregate` operator can be used to sum up the values of a given attribute or set of attributes for the tuples in the window; this operator also relies on a windowing option to group together a set of tuples to address the same challenges outlined in the "Sort" section. An `aggregate` operator also allows for `groupBy` and `partitionBy` parameters to divide up the tuples in a window and perform an aggregation on those smaller subsets of tuples. You can use the `aggregate` operator to perform `count`, `sum`, `average`, `max`, `min`, `first`, `last`, `count distinct`, and other forms of aggregation.

Beacon

A `beacon` is a useful operator because it's used to create tuples on the fly. For example, you can set up a `beacon` to send tuples into a stream at various intervals defined either by a time period (send a tuple every n tenths of a second) and/or by iteration (send out n tuples and then stop). The `beacon` operator can be useful in testing and debugging your Streams applications.

Throttle and Delay

Two other useful operators can help you manipulate the timing and flow of a given stream: throttle and delay. The throttle operator helps you to set the "pace" of the data flowing through a stream. For example, tuples that are arriving sporadically can be sent to the output of the throttle operator at a specified rate (as defined by tuples per second). Similarly, the delay operator can be used to change the timing of the stream. A delay can be set up simply to output tuples after a specific delay period; however, with delay, the tuples exit the operator with the same time interval that existed between the tuples when they arrived. That is, if tuple A arrives 10 seconds before tuple B, which arrives 3 seconds before tuple C, then the delay operator will maintain this timing between tuples on exit, after the tuples have been delayed by the specified amount of time.

Split and Union

The split operator will take one input stream and, as the name suggests, split that stream into multiple output streams. This operator takes a parameterized list of values for a given attribute in the tuple and matches the tuple's attribute with this list to determine on which output stream the tuple will be sent out. The union operator acts in reverse: it takes multiple input streams and combines all the tuples that are found in the input streams into an output stream.

Streams Toolkits

In addition to the adapters and operators described previously, Streams also ships with a number of rich toolkits that allow for even faster application development. These toolkits allow you to connect to specific data sources and manipulate data that is commonly found in databases, financial markets, and much more. Because the Streams toolkits can accelerate your time to analysis with Streams, we figure it's prudent to spend a little time covering them here in more detail; specifically, we'll discuss the Database Toolkit and the Financial Markets Toolkit section.

The Database Toolkit: Operators for Relational Databases

The Database Toolkit allows a stream to read or write to an ODBC database or from a SolidDB database. This allows a stream to query an external database to add data or verify data in the stream for further analysis. The operators available in this Streams toolkit include the following:

`ODBCAppend`	Inserts data into a table from a stream using SQL `INSERT` commands
`ODBCEnrich`	Reads data from a table and combines it with the tuples in the stream
`ODBCSource`	Reads data from a table and puts each row into the stream as a tuple
`SolidDBEnrich`	Reads data from a SolidDB table and adds that information to tuples in the stream

Financial Markets Toolkit

The Financial Information eXchange (FIX) protocol is the standard for the interchange of data to and from financial markets. This standard defines the data formats for the exchange of information related to securities transactions. The Streams Financial Markets Toolkit provides a number of FIX protocol adapters such as:

`FIXMessageToStream`	Converts a FIX message to a stream tuple
`StreamToFIXMessage`	Formats a stream tuple into a valid FIX message for transmission

In addition to these operators, other useful components such as market simulation adapters to simulate market quotes, trades, orders, and more are provided with this toolkit. It also includes adapters for WebSphere MQ messages and WebSphere Front Office for financial markets. All in all, this toolkit greatly reduces the time it takes to develop, test, and deploy stream processes for analyzing financial-based market data.

Enterprise Class

Many real-time application and parallel processing environments built in the past have come and gone; what makes Streams so different is its enterprise class architecture and runtime environment which are powerful and robust enough to handle the most demanding streaming workloads. This is the value that IBM and its research and development arms bring to the Big Data problem. Although some companies have massive IT budgets to try and do this themselves, wouldn't it make sense to invest those budgets in core competencies and the business?

Large, massively parallel jobs have unique availability requirements because in a large cluster, there are bound to be failures. The good news is that Streams has built-in availability characteristics that take this into account. Also consider that in a massive cluster, the creation, visualization, and monitoring of your applications is a critical success factor in keeping your management costs low (as well as the reputation of your business high). Not to worry: Streams has this area covered, too. Finally, the integration with the rest of your enterprise architecture is essential to building a holistic solution rather than a stove pipe or single siloed application. It's a recurring theme we talk about in this book: *IBM offers a Big Data platform, not a Big Data product.*

In this section, we cover some of the enterprise aspects of the Big Data problem for streaming analytics: availability, ease of use, and integration.

High Availability

When you configure your Streams platform, you tell a stream which hosts (servers) will be part of the Streams instance. You can specify three types of hosts for each server in your platform:

- An *application host* is a server that runs SPL jobs.

- A *management host* runs the management services that control the flow of SPL jobs (but doesn't explicitly run any SPL jobs directly), manages security within the instance, monitors any running jobs, and so on.

- A *mixed host* can run both SPL jobs and management tasks.

In a typical environment you would have one management host and the remainder of your servers would be used as application hosts.

When you execute a streaming application, the processing elements (PEs) can each execute on a different server, because, quite simply, PEs are essentially the operators and adapters that make up your streaming application. For example, a source operator can run on one server, which would then stream tuples to another server running operator A, which could then stream tuples to another server running operator B. The operator on this last server would then stream tuples to the sink operator running on yet another server.

In the event of a PE failure, Streams will automatically detect the failure and take any possible remediation actions. For example, if the PE is restartable and relocatable, the Streams runtime will automatically pick an available host on which to run the job and start that PE on that host (and "rewire" the inputs and outputs to other servers as appropriate). However, if the PE continues to fail over and over again (perhaps due to a recurring underlying hardware issue), a retry threshold indicates that after that number of retries is met, the PE will be placed into a `stopped` state and will require manual intervention to resolve the issue. If the PE is restartable, but has been defined as not relocatable (for example, the PE is a sink that requires it to be run on a specific host), the Streams runtime will automatically attempt to restart the PE on the same host, if it is available. Likewise, if a management host fails, you can have the management function restarted elsewhere, assuming you have configured the system with `RecoveryMode=ON`. In this case, the recovery database will have stored the necessary information to restart the management tasks on another server in the cluster.

Consumability: Making the Platform Easy to Use

Usability means deployability. Streams comes with an Eclipse-based visual toolset called *InfoSphere Streams Studio (Streams Studio)*, which allows you to create, edit, test, debug, run, and even visualize a Streams graph model and your SPL applications. Much like other Eclipse-based application development add-ins, Streams Studio has a Streams perspective which includes a Streams Explorer to manage Streams development projects. The Streams perspective also includes a graphical view that lets you visualize the stream graph from one or more sources to one or more sinks and lets you manipulate the graph to manage the application topology.

When you are running an SPL application, Streams Studio provides a great deal of added benefits. Built-in metrics allow you to view the streaming application to surface key runtime characteristics such as the number of tuples in and out of each operator, and more. A log viewer lets you view the various logs on each of the Streams cluster's servers, and an interactive debugger lets you test and debug your applications.

If you click a Streams operator in Streams Studio it opens the SPL editor for that specific operator, which is infused with all sorts of syntax and semantic-related items that make coding the task at hand easier as it steps you through the development process. Finally, there's an integrated help engine that comes in awfully handy when you're developing, debugging, and deploying your Streams applications. All in all, Streams Studio offers the ease of use you would expect from a feature-rich application integrated development environment (IDE) that is part of the full Streams platform, rather than just the parallel execution platform that other vendors offer.

Integration is the Apex of Enterprise Class Analysis

The final aspect of an enterprise class solution is how well it integrates into your existing enterprise architecture. As we've discussed previously, Big Data is not a replacement for your traditional systems; it's there to augment them. Coordinating your traditional and new age Big Data processes takes a vendor that understands both sides of the equation. As you've likely deduced after reading this chapter, Streams already has extensive connection capability into enterprise assets, such as relational databases, in-memory databases, WebSphere queues, and more.

In the preceding section, we briefly talked about Streams' Eclipse-based IDE plug-in and monitoring infrastructure, which allows it to fit into existing application development environments such as Rational or other toolsets based on the widespread de facto standard open source Eclipse framework (which IBM invented and donated to open source, we might add). But that's just the beginning: Streams has sink adapters that allow streaming data to be put into a BigInsights Hadoop environment with a high-speed parallel loader for very fast delivery of streaming data into BigInsights (via the BigInsights Toolkit for Streams) or directly into your data warehouse for your data-at-rest analysis.

As we've talked about throughout this book, Big Data problems require the analysis of data at rest and data in motion, and the integration of Streams and BigInsights offers a platform (not just products) for the analysis of data in real time, as well as the analysis of vast amounts of data at rest for complex analytical workloads. IBM gives you the best of both worlds, and it is brought together under one umbrella with considerations for security, enterprise service level agreement expectations, nationalization of the product and support channels, enterprise performance expectations, and more.

Additional Skills Resources

Rely on the wide range of IBM experts, programs, and services that are available to help you take your Big Data skills to the next level. Participate in our online community through the BigInsights wiki. Find whitepapers, videos, demos, download of BigInsights, links to twitter, blog and facebook sites, all the latest news and more.

Visit **ibm.com/**developerworks/wiki/biginsights

IBM Certification & Mastery Exams

Find industry-leading professional certification and mastery exams. New mastery exams are now available for BigInsights (M97) and InfoSphere Streams (N08).

Visit **ibm.com**/certify/mastery_tests

IBM Training

Find greener and more cost-effective online learning, traditional classroom, private online training, and world-class instructors. New classes added frequently and in a variety of formats.

Visit **ibm.com**/software/data/education to check out available education courses.

- InfoSphere BigInsights Essentials using Apache Hadoop
- Analytics Fundamentals with BigInsights – Part 1
- Programming for InfoSphere Streams
- Administration of InfoSphere Streams v2

Information Management Bookstore

Find the electronic version of this book, links to the most informative Information Management books on the market, along with valuable links and offers to save you money and enhance your skills.

Visit **ibm.com**/software/data/education/bookstore